M000206411

Thinking Like a Phage

The Genius of the Viruses That Infect
Bacteria and Archaea

by Merry Youle

Illustrations by Leah Pantéa

San Diego, CA

First edition, May 2017

Copyright © 2017 by Merry Youle

Illustrations by Leah Pantéa

Thinking Like a Phage by Merry Youle is licensed under a Creative Commons Attribution-NonCommercial-NoDerivs 4.0 International License which permits non-commercial use without modification provided author and source are credited.

Library of Congress Cataloging-in-Publication Data
Youle, Merry
Thinking like a phage: the genius of the viruses that infect Bacteria and Archaea
Includes bibliographical references.
Library of Congress Control Number: 2017901174

ISBN: 978-0-9904943-1-7

Book design by Alexis Morrison, Set Right Typography

Published by Wholon, San Diego, CA

Printed in the United States of America

Dedication

To Forest Rohwer and Elio Schaechter

who, each in their own way, handed me a microphone

and encouraged me to use it

Table of Contents

    in which we meet the 21 pheatured phages who
    star in the chapters that follow.

    in which phage chromosomes, having entered their
    host cell, outwit the patrolling cellular defenses.

    in which the phages take over their host cell and
    redirect its activities toward making more phages.

    in which the newly-minted phage proteins and
    chromosomes assemble themselves into progeny
    virions.

    in which the progeny virions escape from the cell
    that had supported their production.

in which lucky drifting virions collide with a
potential host cell and locate their receptor on its
surface.

in which virions deliver their chromosome into
a new host cell – the first step in launching an
infection.

in which temperate phages reside for extended
periods within their host cell for mutual benefit.

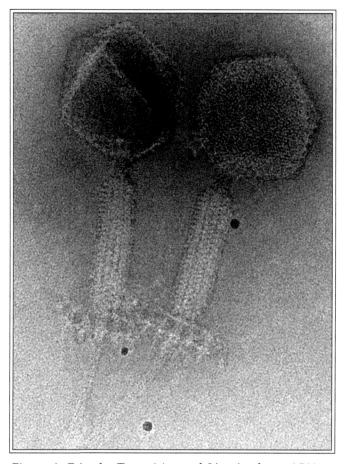

Figure 1: Friends. Two virions of *Listeria* phage A511, a myovirus, one shown before (right) and one after (left) ejection of its DNA chromosome. Courtesy of Jochen Klumpp, ETH, Zurich, Switzerland.

Preface

This book was born of love, exasperation, and wonderment.

While in grad school in the late 1960s, I fell in love. That's when I was first introduced to the bacteriophages. At that time, phages were curiosities that were discussed primarily as useful tools for studying bacterial genetics. Electron micrographs of the classic phage silhouette – icosahedral head, contractile tail, spidery tail fibers – suggested there was more to their story. When I returned to biology after a hiatus of more than three decades, I found that an astounding amount had been learned about them. Phages were more than I had imagined – more numerous, more diverse, more skillful in their way of life, more influential in every ecosystem. I was enthralled. I read on and started writing their stories soon thereafter.

Exasperation arose, incited by the unrelenting vilification of "my" phages. To this day, research funding, virology textbooks, and popular science writing emphasize the viral minority responsible for disease in us or our domesticates. Smallpox, measles, flu, and HIV, to name a few, have indeed had immense impact on human history. For myself, I am reminded daily of my childhood encounter with polio virus. While it is prudent for us to understand these viral pathogens so we can more wisely safeguard our health, not every virus is a bad virus.

The phages were painted with the same brush, or simply ignored. It was easy to discount them because they did not appear to be numerous enough in the environment to matter. Now, with improved tools, we can see a planet alive with phages. They are the most numerous and most diverse life form on Earth. Every ecosystem rests upon vast numbers of them and their prokaryote hosts; together they are the foundations of both ecology and evolution. We also see phages as potential allies in our efforts to prevent and fight bacterial infections – efforts known as phage therapy. In addition, phage virion components are being co-opted for drug delivery and nanotechnology applications. And, most intimately, phages contribute to the healthy microbiome essential for our individual well being. So, rather suddenly, phages are front and center, and cast in a beneficent light.

I have taken an anthropomorphic approach in this book, even though such a tone should be used with caution in scientific discourse. The title is not meant to imply that phages think, either individually or collectively, but to engage your imagination. Pause for a moment and reflect on the challenges you would face as a phage. Appreciate the varied and elegant solutions they employ. Their individual stories are the grist for this book. The book to follow, *Phage in Community*, will explore the complexities of phage life in the real world and the essential roles of the phages in the ecology and evolution of all life on Earth.

The wonderment? There were three discoveries in the past century that revealed vast, previously unknown realms. The recognition of the phage world was one of them. The other two? The very small (quantum field theory and particle physics) and the very large (the expanse of space-time). All three of these discoveries are unsettling to those who were comfortable with our being at the center of the universe. Homo sapiens has been progressively displaced. No longer the superstar, we find ourselves relegated to a minor walk-on role in a small theatre in the boondocks. Should we be dispirited? I think not. We and the phages are both part of a far grander drama, one that we can barely imagine and cannot fully comprehend. There was wonder aplenty to be experienced by simply looking around us at the familiar macrobes – the plants and animals. That wonder was magnified in school when we watched a paramecium swim purposely about within a drop of water or peered through a telescope at the rings of Saturn. To that, we have now added electron micrographs of the faceted gems of the phage – static images from another dimension that inspire even greater astonishment. Although outside our normal vision, the phage multitude is always there – an ancient, dynamic, bubbling, creative force that underpins all life on Earth.

I invite you to explore the world of the phages, and to wonder, with me, about this world we share.

Merry Youle
February, 2017
Ocean View, Hawaii

Acknowledgements

First to be acknowledged are the two people whose years of steadfast generosity and encouragement made it possible for me to undertake this book project. In 2007, Elio Schaechter welcomed me into his new-born microbiology blog, *Small Things Considered*. During the years that followed, he shared the helm with me and published more than 80 of my posts, most of them about the phages. In 2008, when I was seeking editing work, he played matchmaker and put me in touch with Forest Rohwer. Forest proceeded to then give me one writing and publishing opportunity after another, culminating in our publication of *Life in Our Phage World* in 2015. En route, I could count on him as an inexhaustible source of creative ideas, mind-stretching challenges, and support. I still shake my head in amazement when I look back over the incredible good fortune these two brought into my life.

The book you are now reading owes its engaging form to the contributions of several other people. Key among them are two who worked by my side during production of Forest's phage book. I knew I wanted them on my team for this one. Leah Pantea, a fine artist located in San Diego, provided all the illustrations – a task that required not only her skill and imagination, but also patient attention to detail as we revisited the fine points of virion structure and phage terminology. Alexis Morrison, a neighbor of mine here on the Big Island, is a graphics wizard with a discerning eye and a keen sense of humor. She designed this book and then made it an appetizing reality. Another veteran of that project, Heather Maughan, assisted me month after month in countless diverse ways. All I had to do was ask. I also must mention Graham Hatfull, a trendsetter for verbal phage phun and the phounder of the PHIRE and SEA-PHAGES programs (see "Other Resources" on page 273). Through those programs, he provided my imagined reader–a student eagerly discovering the phage world, including isolating a phage of their very own. This book is especially for them.

Several people read through draft chapters, each of whom contributed unique insights from their particular perspective. The book was much improved as a result. These included Bentley Fane, Nadine Fornelos,

Jamie Henzy, Breeann Kirby, Heather Maughan, Sally Molloy, Forest Rohwer, Elio Schaechter, and Christoph Weigel.

I gratefully acknowledge the generosity of the many members of the phage community who freely contributed images unencumbered by copyright. Each donor is noted in the legend accompanying each figure, but they deserve more of an acknowledgement. For many of them, fulfilling my image request required more than a few mouse clicks. Sometimes it meant searching through dusty archives, even those located at a former campus home. One packaged and sent me an old negative by snail mail, while others generated new tomographic reconstructions so that I could have an unpublished version. The net result is a library of images worthy of the phages.

It summary, it takes a community to create a phage book. It also takes the phages to create a community.

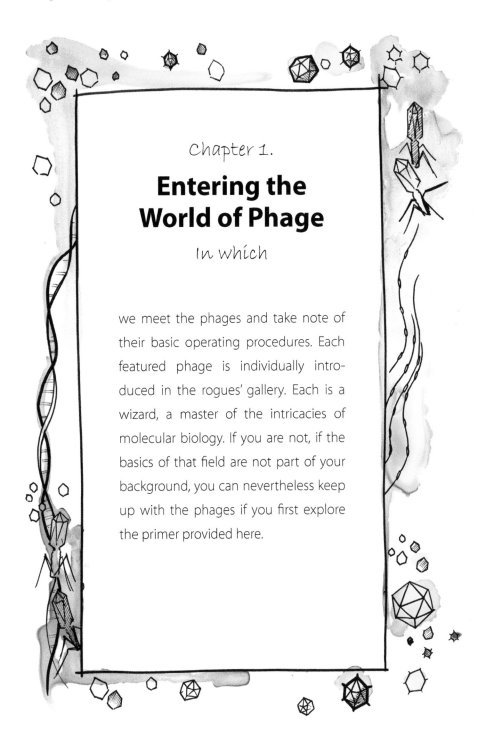

Chapter 1.

Entering the World of Phage

In which

we meet the phages and take note of their basic operating procedures. Each featured phage is individually introduced in the rogues' gallery. Each is a wizard, a master of the intricacies of molecular biology. If you are not, if the basics of that field are not part of your background, you can nevertheless keep up with the phages if you first explore the primer provided here.

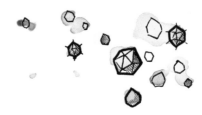

...it has been well said that a virus is "a piece of bad news wrapped up in protein."
P.B. Medawar and J.S. Medawar, 1985

Whether or not viruses should be regarded as organisms is a matter of taste.
André Lwoff, 1962

Nevertheless, I believe that the virus factory should be considered the actual virus organism when referring to a virus.
Jean-Michel Claverie, 2006

Viruses are by common definition neither organisms nor alive.
Harald Bruüssow, 2009

Because, after all, a person's a person, no matter how small.
Dr. Seuss, *Horton Hears a Who*

This is biology, so there are of course exceptions to any rules we might attempt to derive.
Sherwood Casjens, 2003

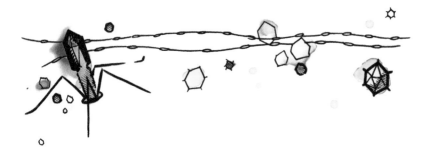

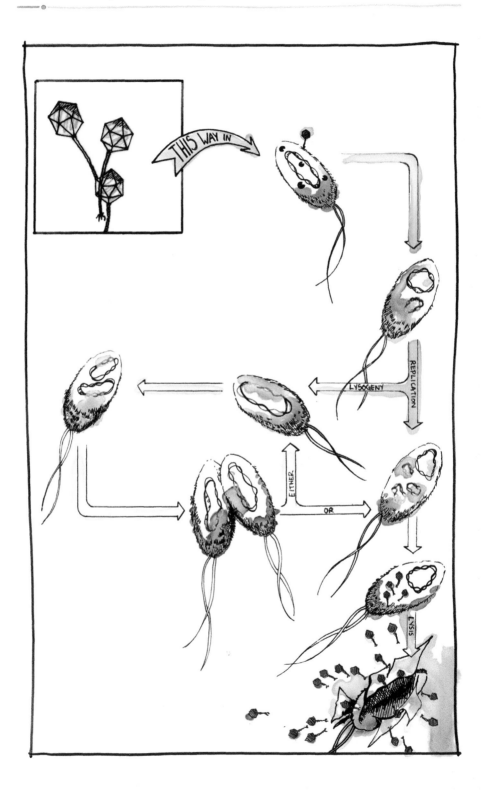

Two Points of View

Bobbing in mid-Pacific waves, or swirling in the roiling waters of an acidic hot spring, stuck in the mucus covering a coral polyp, ensconced in a crack in a desert rock, or trapped in the remains of an enchilada in your gut, a bacterium bursts open and 25, a hundred, perhaps a thousand multi-faceted jewels spill out from the carcass—the fruits of a successful phage infection. Riding the currents or ricocheting off cells and flotsam, these tiny hopeful particles disperse randomly, drifting. Occasionally a lucky one collides with a suitable bacterium and sticks. Then minutes, hours, days, or months later, this cell, too, ruptures and another horde ventures forth. The odds are against individual success, but the voyagers are many. If fortune smiles, one of them will arrive at an obliging door, will enter and dine, and will repeat this ancient cycle yet once again.

...

It's a bacterium's worst nightmare, the arrival of that one phage in a million who will evade all of its state-of-the-art defenses, who moreover arrives with a plan of its own, speaks the language understood by cellular machinery, and knows how to divert the cell's energy and resources to virion production. This particular bacterium's fate now is to support the multiplication of enemy troops and ultimately to release them into the neighborhood to prey upon its kin. Is its lineage doomed? It is, after all, out-numbered ten-to-one, and its family risks being out-maneuvered by the flagrant fecundity and rapid innovation of the phages. Not being so easily outdone, Bacteria meet the phage challenge again and again with innovations of their own, regaining the upper hand once more. Moreover, they can take some comfort in reminding themselves that the phages need them. Without hosts, the phages are at a dead end. Phage and host have co-existed for billions of years, the temporary advantage seesawing back and forth. Why worry? The cell continues to prepare to give birth to two daughters.

Whose side are you on?

What is a phage? That's easy. It is a virus that infects a prokaryote. But now, instead of only one, we have three words to define: *virus*, *infect*, and *prokaryote*.

Virus is, at its root, a misnomer derived from the Latin *vīrus*, a poisonous liquid. An accurate and enduring definition was provided in 1978 by Salvador Luria, one of the key members of the phage group:[1] *Viruses are entities whose genomes are elements of nucleic acid that replicate inside living cells using the cellular synthetic machinery and causing the synthesis of specialized elements that can transfer the viral genome to other cells.* Viruses are not the only entities now known to meet this definition. Therefore, this is often qualified further by noting that viral chromosomes[2] travel from cell to cell enclosed in a protein shell, or capsid.

Infect refers to the entry of a viral chromosome into a living cell that it then manipulates to support its own replication.

Prokaryote is shorthand for the organisms that compose two of the three branches on the Tree of Life (see Figure 2). These two branches, originally known as the Bacteria and the Archaebacteria, were initially combined into one supergroup, the prokaryotes, because both groups are single-celled organisms without membrane-bounded intracellular compartments. Thus, diverse organisms were classified together based on what they lacked – a true nucleus and the other organelles found in all eukaryote cells. Genome sequencing later demonstrated marked differences in the cellular machinery and evolutionary histories of these two prokaryote branches. The Archaebacteria, renamed the Archaea, were then recognized as being a third distinct domain[3] of life, as different from the Bacteria as either group is from the Eukarya.[4]

---

[1] The phage group was an informal network of researchers in the USA that laid the foundations in the mid-20th century for the understanding of phage biology. Their work also developed phages as tools for the new field of molecular biology.

[2] Throughout this book, I use *chromosome* to refer to the molecule(s) of nucleic acid in which genetic information is encoded, and reserve the term *genome* for discussion of the genetic information itself.

[3] domain: one of the three major divisions of cellular life: Bacteria, Archaea, Eukarya.

[4] This three domain view has recently been challenged by a two domain model, the two domains being the Bacteria and the Archaea. It has been convincingly argued on the basis of genomic analyses that eukaryotes emerged from within the archaeal domain. See "Further Reading" on page 59 for references.

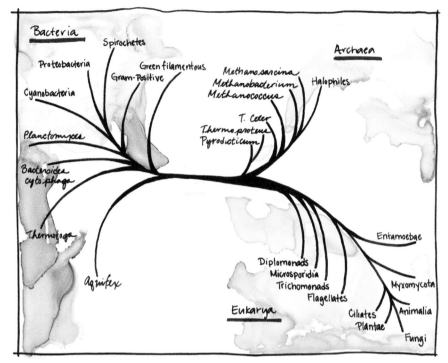

Figure 2: Tree of Life

This definition of a phage is arguably correct, but it does not convey what makes the phages so fascinating, so intriguing, and so important. That is the job of the rest of this book and its planned sequel.

It has long been debated whether or not viruses are alive. The answer depends on how you define *life*. A related argument flourishes today, this one questioning whether or not viruses should be included in the "universal" Tree of Life.[5] This tree portrays family relationships among cellular organisms all of which share a common cellular ancestor. As such, there is no place on it for any viruses. However, because it ignores all viruses, such a tree cannot portray the full story of the evolution of life on Earth. Viruses are the most numerous and most diverse life forms. They evolve genes, they move genes between organisms, and they profoundly affect the diversity and metabolic activity of every ecosystem. I'd venture to say that wherever "life" exists in the universe, virus-like entities will abound. There, too, they will be essen-

---

[5] See "Further Reading" on page 59 for relevant papers.

Figure 3: Seeing virions. (Left) Helical capsid of tobacco mosaic virus (TMV), the first virions observed under the electron microscope. Transmission electron micrograph (TEM) using rotary shadowing with platinum courtesy of Marcus Drechsler, University of Bayreuth, Germany. (Right) The elaborate virions of Lander, a member of the *Myoviridae*, the phage family whose virions possess a long contractile tail attached to an icosahedral capsid. Courtesy of Jun Liu, University of Texas Health Sciences Center, Houston.

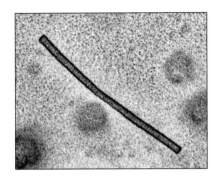

tial to life's evolution and ecology. Viruses are an integral part of our living universe (see PIC[6]).

Discovery

A century ago, the bacterial pathogens responsible for numerous infectious diseases had been successfully isolated, cultured, and characterized. These entities could be removed from liquid culture media by passage through Chamberland or Berkefeld ceramic filters. However, the infectious agents responsible for some diseases passed right through the 0.1 to 1.0 μm filter pores. Whatever these mysterious things were, they were too small to be Bacteria. These "poisons" or "toxic liquids" were called viruses.[7] Subsequent work provided evidence that viruses were, after all, particles – very small particles. The invention of the electron microscope (EM)[8] in 1931 made it possible a few years later to put a face to the name, and what a stunning face it was! Under the EM viruses showed themselves to be particles with complex geometric structures, present in millions of copies identical in size, shape, and detail.

---

[6] *Phage in Community*, the book planned to follow this one, will explore phage ecology, evolution, competition, and other collective activities.

[7] In 1898, M. W. Beijerinck applied the Latin descriptor *contagium vivum fluidum* to the mysterious agent causing a mosaic pattern on the leaves of tobacco and many other plants. This agent was later known as tobacco mosaic virus (TMV).

[8] Electron microscope: a microscope that images small objects using a beam of electrons rather than visible light to achieve the higher magnification and resolution required to visualize fine details within nanometer-scale biological structures, including virions.

The first virus particles seen were those of a plant virus, not a phage (see Figure 3). These were the simple, rod-like agents that cause tobacco mosaic disease (tobacco mosaic virus). Today people routinely refer to isolated particles such as these as viruses or phages. I reserve the terms *virus* and *phage* for the active entities residing inside a cell. The particles, or virions,[9] are akin to seeds and spores. Like them, virions are metabolically inert units released for dispersal. They cannot generate energy to power their movement or carry out any other activity. A virion can only drift among the flotsam, its fate resting on a chance collision with a potential host cell. Only if delivered into a suitable host can the phage chromosome within spring into action. As with plant seeds, a great number of virions are produced so that at least one will survive and reproduce.

Usually, but not Always

The term *phage* refers, *sensu stricto*, to viruses that "eat" Bacteria, i.e., the bacteriophages. When thinking like a phage, I also include a few viruses that infect Archaea (Archaebacteria). Because we know so much more about the viruses that infect Bacteria, these pages will usually, but not always, be focused on them.

However you define them, phages differ widely among themselves in their properties and their strategies. There are extremely few statements that one can make that apply to all of them. It would surely be tiresome to read – or to write – again and again, qualifiers such as *some, most, typically, usually, almost always, with few exceptions, the majority,* and so on. I have chosen instead to usually, but not always, forego the qualifiers when relating that which is most commonly observed. Please take all generalizations in this book with the proverbial grain of salt. For every phage rule there are numerous inventive exceptions.

---

[9] virion: the inert, extracellular dispersal form of a phage composed of the phage chromosome(s) inside a protein shell (capsid) and, sometimes, also a tail, tail fibers, and/or a lipid membrane.

Earning, or Stealing, a Living

Packaged inside each virion is the chromosome that carries the genetic information that makes this phage a particular type of phage. This genomic cargo is shielded from environmental insults while in transit by a multi-faceted protein shell, its capsid. A capsid is more than a resilient shipping carton. It also recognizes a host when it collides with one. First contact is made with the outermost layer of the cell envelope or with a structure, such as a flagellum[10] or a pilus,[11] that extends outward beyond the envelope. If the virion encounters its specific receptor[12] there, it binds (adsorbs) and then delivers the chromosome through the cell membrane and into the cytoplasm. Afterwards it cleans up after itself by resealing the punctured membrane—virion mission accomplished. There is no longer a phage and a bacterium, but rather a virocell[13] with a phage chromosome actively influencing cellular activities. Who is in charge now?

It depends. The best known phage scenario is the archetypal alien takeover leading shortly to the death of the cell. The cell is enslaved forthwith, but it is not killed immediately. Its continued activity is required to provide energy and other essential services for phage replication. The virocell may continue to swim merrily along while its metabolism proceeds in part under phage orders. Virion components are needed — proteins for the capsid, chromosomes to be packaged inside. The cell's protein synthesizing machinery is redirected to manufacture phage proteins, while the phage chromosome is replicated again and again. Some thrifty phages chop the host chromosome into bits to be recycled for production of more phage chromosomes. Completed virions accumulate quietly inside the cell. At the right time, the phage triggers the explosive rupture of the cell (lysis[14]). Virocell "guts"

---

[10] flagellum (plural, flagella): an appendage of many prokaryote cells that is anchored in the cell membrane and rotates to propel the cell through the milieu.

[11] pilus (plural, pili): straight, filamentous appendages of some prokaryote cells that carry out one of a variety of functions such as attachment to host cells or surfaces, transfer of DNA between cells, or motility.

[12] receptor: the molecular structure on the surface of a cell to which a phage virion specifically attaches, reversibly or irreversibly, as the first step in infection.

[13] virocell: a host cell that contains an intact phage chromosome that is capable of replication and virion production inside this cell.

[14] lysis: the rupture of a prokaryote cell membrane that results when a weakened or damaged cell wall is unable to counter the internal turgor pressure of the cell.

and progeny virions spew out. The virions are on their way in search of new hosts, there to repeat this ancient cycle.

Unseen by you or me, this deadly sequence is initiated about 10^{25} times every second of every day, and has been happening for billions of years.[15] Such numbers are unimaginable. We can easily talk about them and manipulate them in calculations, but can you really picture 10^{25} of anything?

Immediate replication followed in short order by host death is not the only successful phage strategy. Some phages, the temperate phages,[16] have the option to instead abide quietly while the virocell grows and divides, generation after generation. In this case, the phage chromosome, now known as a prophage,[17] usually integrates into the host chromosome where the virocell will replicate it along with its own DNA. Since a prophage is inherited by each of the daughter cells, prophages double in number along with the virocell. As rent, the prophage provides services such as blocking infection by related phages. It often happens that at some later time, perhaps triggered by significant damage to the virocell's DNA, the phage breaks its lease and switches to the lytic pathway. Now it quickly replicates and, when its progeny virions are ready to exit, lyses the virocell.

A third option, pursued by only a few phage types, is to convert the host into an ongoing virion production facility. Lysis is abolished. Phage and host apportion resources. The host continues to grow, albeit more slowly. Virions extrude through the cell membrane while the enslaved cell lives on to support continued virion production.

Behind every successful phage infection is a fine-tuned program of host exploitation. Given gazillions of phages evolving for billions of years, the phages have had the opportunity to experiment with every imaginable possibility for every step in their life cycle, as well as many variations we could never have imagined. An infection begins with

---

[15] The number 10^{25} is scientific notation representing the number 1 followed by 25 zeros (10,000,000,000,000,000,000,000,000). Scientific notation is used throughout this book.

[16] temperate phage: a phage that is capable of two modes of infection, i.e., immediate lytic replication and lysogeny.

[17] prophage: a phage chromosome that is not actively replicating but is maintained stably in a virocell, usually integrated into the cell's chromosome.

the delivery of a phage chromosome into a productive workshop currently being managed for maximum cell growth under the existing conditions. Comprehensive cellular infrastructure is already in place. The cell possesses a means of energy production, efficient methods for raw material procurement, organized metabolic pathways, enzymatic tools, equipment for protein synthesis, and much more. The successful phage takes charge and redirects these activities toward virion production, without compromising needed virocell capabilities in the short term. This it does skillfully, sometimes by mimicking or exploiting the cell's own regulatory signals. Specific cellular components are recruited for particular services. So intimate is this dance that each phage can infect at most only a few related bacterial or archaeal species, often only one species, and sometimes only one specific strain within that species. Conversely, every prokaryote cell type is host to at least one phage, perhaps ten or more different types. These phages can be members of different phage families (see "Bacteriophage Families" on page 14), may differ in virion morphology and chromosome type, and may use conspicuously different infection strategies. But they do have one thing in common: sophisticated tactics for replicating within this particular prokaryote host.

Vive la Différence!

A living cell, even a "simple" prokaryotic cell, is a teeming, organized, structured hive of activity. Although these cells lack internal membrane-bounded compartments such as a nucleus, the stuff of the cell is not randomly scattered throughout the cytoplasm. The highly-condensed DNA, along with enzymes for its replication and for gene expression, are localized in a distinct region (the nucleoid). For some metabolic pathways, the various enzymes required to carry out the sequential steps are housed together within protein-bounded nanocompartments (see "A Two-Way Street" on page 153). Cell growth and division, as well as the maintenance of cell shape, depend on structures formed by the polymerization and depolymerization of protein filaments. The cell membrane and the additional protective layers surrounding it are structurally and chemically complex. Many other proteins are found in specific cell regions, including proteins required by the replicating phage as well as some that are poised to destroy an invading phage chromosome.

Omnis cellula e cellula.[18] Every cell arises from a pre-existing cell. None are formed *de novo.* The mother cell divides into two cells that, although sometimes unequal in size, both receive a copy of the cell's chromosome(s) and some cytoplasm bounded by a cell membrane. Thus, a daughter cell inherits not only genes from its parent, but also cellular structure and a functioning metabolism. It hits the ground running. In contrast, a phage inherits only the information encoded in a DNA or RNA molecule, plus a few proteins delivered along with the chromosome for immediate use. The parental virion shell remains outside the cell. If a human introduces a naked phage chromosome into a potential host cell, the phage can still replicate and produce fully competent progeny. Given modern technology, you could take this one step farther and dispense altogether with the parental chromosome. How? You could sequence a phage genome, email the sequence to a friend a thousand miles away, and have her synthesize a DNA molecule with the same sequence. This molecule that never saw a mother phage can, when introduced into a host cell, replicate and oversee the construction of infectious virions. The essence of a phage is the information encoded by its genome (discussed further in PIC). No matter how large the phage genome, no matter how many proteins it encodes, a phage is not a cell. The information it carries provides instructions that require the cellular machinery for implementation. A phage can replicate only if it can exploit the structures and metabolism of a suitable living cell.

Classification by Virion and Chromosome Type

In the decades following the development of the electron microscope, researchers examined the virions of more and more viruses. Some architectural themes were repeated again and again, accompanied by intriguing variations on those themes. Given that all viruses make virions, and given the long tradition of classifying multicellular organisms based on morphological traits, it seemed natural to classify phages into families based on virion similarities. Viruses also offer the option to classify by another virion trait: the molecular structure of the chromosome packaged inside. All cells and many phages use double-stranded molecules of DNA for their chromosome, but some phages

---

[18] Rudolf Ludwig Karl Virchow 1859

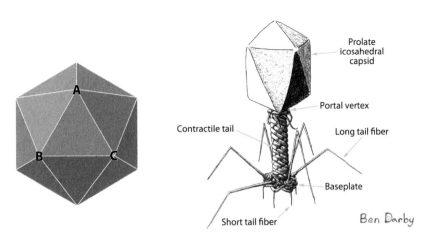

Figure 4: Phage virion architectures. (Left) An icosahedron, with twenty triangular faces, twelve vertices, and 30 edges. A, B, and C mark vertices; AB, BC, and AC denote edges; and the triangle ABC comprises one face. The axes of rotational symmetry include a three-fold axis in the center of each face, a five-fold axis at each vertex, and a two-fold axis at each edge. Courtesy of Kenneth J. M. MacLean. (Right) Drawing of the virion architecture of *Synechococcus* phage S-SSM7, a large myovirus (family *Myoviridae*) similar to Lander (see below). Original drawing by Ben Darby. Previously published in *Life in Our Phage World* by Rohwer, F, M Youle, H Maughan, N Hisakawa. 2014. Wholon. Used with permission.

use RNA. During transport between cells, many phage chromosomes are linear, but some are circular. Similarly, many are composed of a double-stranded molecule of nucleic acid, but some make do with but a single-strand. Virion morphology and chromosome type form the basis of viral taxonomies still in use today.[19]

The year 1977 saw the sequencing, with great effort, of the first DNA-based genome. For this project, the researchers chose the smallest DNA genome then known – Yoda's genome of only 5386 nucleotides (see "Yoda" on page 41). Sequencing technologies developed since then have reduced the labor and the cost by many orders of magnitude, thereby enabling the highly publicized sequencing of the 3 billion base pair human genome. More quietly, several thousand phage genomes have been sequenced which, although a very small and biased sample of the extant phages, has prompted the development of several clas-

---

[19] The International Committee on Taxonomy of Viruses (ICTV) system (http://bit.ly/2d53CAH). Also, the Baltimore system classifies viruses based solely on chromosome type (http://bit.ly/2cW7RcN).

sification systems based on genome sequence.[20] Phage classification remains unsettled due to factors inherent in phage evolution (see PIC).

The great majority of known phages that infect Bacteria share the same basic virion design: an icosahedral[21] capsid enclosing a double-stranded DNA chromosome and bearing a tail attached at one of the twelve vertices (see Figure 4). The obviously different tail types led to the assignment of these phages to three families within the order Caudovirales (tailed viruses): the siphoviruses (*Siphoviridae*) with a long, flexible tail; the myoviruses (*Myoviridae*) with a long contractile tail; and the podoviruses (*Podoviridae*) with a short stubby tail. Long or short, all of these tails are complex, precisely-detailed structures that carry out multiple functions during infection. To call them nanomachines, as is often done, is to disparage them, because we have yet to manufacture their equal.

Other bacteriophages find a modest icosahedron to be quite adequate (*Leviviridae, Microviridae*). Moreover, such a simple virion is doable by those with genomes too small to encode the many additional proteins needed for a tail. Some phages with icosahedral virions and larger genomes invest additional genes in a protein-rich lipid membrane that lines the inside (*Corticoviridae, Tectiviridae*) or envelopes the outside (*Cystoviridae*) of the protein capsid. A completely different architecture is used by the filamentous phages (*Inoviridae*) whose virions, as their name implies, are long, slender filaments. Lastly, a phage[22] whose bacterial hosts[23] lack a cell wall manages without a protein capsid and protects its chromosome with only a protein-rich membrane.

Bacteriophage Families

Corticoviridae (Cortic, crust or bark)
    icosahedron; internal membrane
    dsDNA,[24] circular

---

[20] See *Phage Classification for the 21st Century* by Daniel C. Nelson in Rohwer et al., 2014.
[21] icosahedron: a geometric solid with twenty flat faces (each face being an equilateral triangle) and twelve vertices.
[22] *Acholeplasma* phage L2.
[23] *Acholeplasma laidlawii* is a member of a class of small Bacteria that lack a cell wall and whose cell membrane contains sterols (the Mollicutes).
[24] dsDNA, dsRNA: double-stranded DNA, double-stranded RNA.

Cystoviridae (*Cysto*, sack or bladder)
  icosahedron; inner and outer capsids; external membrane
  dsRNA, segmented[25]
Inoviridae (*Ino*, fiber)
  filament
  ssDNA,[26] circular
Leviviridae (*Levi*, light in weight)
  small icosahedron
  ssRNA
Microviridae (*Micro*, small)
  small icosahedron
  ssDNA, circular
Myoviridae (*Myo*, muscle)
  icosahedron, long contractile tail
  dsDNA
Podoviridae (*Podos*, foot)
  icosahedron, short tail
  dsDNA
Siphoviridae (*Sipho*, tube)
  icosahedron, long flexible tail
  dsDNA
Tectiviridae (*Tecti*, covered)
  icosahedron, internal membrane
  dsDNA

The architectural innovations of the archaeal phages are quite striking in comparison. So far 15 families of archaeal phages have been proposed based on their diverse virion architectures, and it is likely that more await discovery. Their repertoire includes the familiar icosahedral capsids with or without tails, flexible or stiff rods, spindles with one, two, or no tails, bottle-shapes, droplet shapes, and others.

Bear in mind that a family can encompass very diverse phages, some of which are only distantly related to one another. They may have markedly different genome sizes, different hosts, and different lifestyles.

---

[25] segmented genome: a genome that is encoded by two or more separate chromosomes.
[26] ssDNA, ssRNA: single-stranded DNA, single-stranded RNA.

Pheatured Phages

We know nothing at all about most phages. Globally, we've sampled only a very small fraction of their total diversity. Many are known only from a bit of genomic sequence recovered from a random sample of DNA from natural communities.[27] Far fewer have been cultured along with their host in the lab, had their genome sequenced, or had their infection tactics observed. Approximately 6,300 had been examined by EM as of 2013. By early 2016, 2,792 bacterial phage genomes and 67 archaeal phage genomes had been sequenced. Even the best studied – the subjects of thousands of published research papers each – still harbor unpredictable surprises. Although the unknown territory is vast, what is known could fill volumes. Of necessity, I have omitted much. Because most research has focused on a handful of "model" phages, you'll find some of these featured again and again throughout these chapters. Overall, I selected stories that illustrate the diverse ways that phages "think." Their "intelligent" strategies are the product of billions of years of evolution, a process that continues today at a speed so rapid that changes are observable in the lab from one day to the next. Many pages in PIC will be dedicated to the relentless evolutionary explorations of the phages and their impact on all life on Earth.

What's in a Name?

When I was first learning about the phages, I was forever losing track of which phage was which. I would confuse T4 with T7, search my memory for the difference between φ6 and φ29, mix up PRD1 and PM2, and so on. You might well be better with names than I, and perhaps this isn't an issue for you. Nevertheless, none of those names help to bridge the chasm that separates their world from ours, nor do they readily bring to mind that phage's distinguishing characteristics. Ergo, I have assigned a

---

[27] virome: a viral metagenome, i.e., a sequence library obtained by sequencing a sample of DNA or RNA extracted from an entire viral community.

nickname to each of the phages you will meet in these pages, a nickname that conveys one of its pertinent characteristics.

Granted, accepted taxonomic nomenclature should be respected and consistently used in scientific writing, but phage nomenclature is in disarray. The common names convey nothing about the phage's properties or host. Worse yet, the same name (e.g., P1) has sometimes been assigned to different phages by different researchers. To rectify this, a logical system of nomenclature was proposed in 2009.[28] However, it has not been widely adopted, partly because its precision also makes it cumbersome. For example, the name for one of the six P1 phages is vB_EcoM-P1 (virus **Bacteria** *E. coli* **Myoviridae-P1**) – clear and informative, but not the sort of name easily dropped into a conversation. Moreover, none of these names help you to recall any of that phage's distinguishing quirks. Lastly, I smile and acknowledge the example provided by the PHIRE and SEA-PHAGES programs (see "Other Resources" on page 273) created by Graham Hatfull. Here the students name the phages that they discover. Their subsequent work has yielded published genome papers for phages such as Flagstaff, Lily, and Corndog.

In the brief introductions below, I have alphabetized the phages by nickname, but also included their common name. Throughout the book, a phage's first appearance in a chapter is typically accompanied by a reminder of its common name. In the index, common names and nicknames are cross-referenced.

Many of the best studied phages infect *E. coli*, the lab rat of the bacterial world, or they infect other closely related enteric[29] Bacteria. As a result, these phages are disproportionately featured here. Coliphages[30] such as Lander, Stubby, Minimalist, and Temperance, appear again and again in the following chapters. Intensive exploration of a few phages has enabled researchers to develop tool sets for working with

---

[28] See Kropinski et al. 2009 in "Further Reading" on page 59.
[29] enteric: (adj.) pertaining to the intestines, often used to denote prokaryotes that are found in the gut of humans or other vertebrates.
[30] coliphage: a phage that infects the bacterium *Escherichia coli* (*E. coli*).

them and thereby to penetrate deeply into some basic cellular activities. On the other hand, this practice has tended to establish paradigms that delay the discovery and appreciation of the full diversity of phage tactics. For example, Bacteria are classified as Gram-positive or Gram-negative based on the way they respond to a particular staining procedure, the Gram stain. Whereas Gram-negative Bacteria are typically enclosed within two membranes separated by a significant moat containing a thin peptidoglycan layer, the Gram-positive ones possess a single membrane surrounded by a thick peptidoglycan cell wall. The challenges faced by a phage during chromosome delivery and later when releasing its progeny virions differ for phages infecting these two bacterial groups. Since most model phages infect Gram-negative Bacteria such as *E. coli*, we know much less about the tactics of phages with Gram-positive hosts.

Some Details

Chromosome sizes are measured in base pairs (bp) or kilobase pairs (kbp) for double-stranded DNA and RNA chromosomes, and in nucleotides (nt) for single-stranded ones. Once a phage's chromosome has been sequenced, the number of genes in its genome can be inferred by various gene prediction computer programs each of which employs a slightly different algorithm or different criteria. Thus, these are the number of predicted, or putative, genes. As methods undergo further refinement, the number of genes thought to be encoded by a particular type of phage may shift slightly.

The phage portraits below include transmission electron micrographs (TEMs)[31] and three-dimensional reconstructions produced using other high-resolution techniques, as well as some artistic, but geometric, interpretations based on these data. All of these approaches enable us to visualize the invisible. We will never be able to see virions with our eyes, limited as our vision is to a narrow range of the electromagnetic spectrum. Observations that rely on sensing visible light cannot re-

---

[31] transmission electron micrograph (TEM): a high resolution image produced by irradiating small particles or ultra-thin slices of a sample with a beam of electrons. The differential absorption of electrons due to the varying composition or structure within the sample is sometimes supplemented by staining or shadowing techniques to increase image contrast.

solve the structure of macromolecules. Thus, we resort to using beams of electrons and X-rays that have shorter wavelengths and are therefore capable of resolving finer detail.[32]

Three-Dimensional Capsid Reconstructions

Current imaging techniques provide high-resolution, three-dimensional reconstructions of icosahedral capsids and procapsids, often to near-atomic resolution. One frequently used approach combines cryo-electron microscopy[33] (cryo-EM) with tomographic reconstruction. For cryo-EM, viral particles, attached to a grid for EM observation, are plunge-frozen at liquid ethane temperatures. Such rapid freezing prevents the formation of ice crystals that would distort or disrupt the viral structure. The sample is held at a temperature below −150° C throughout the imaging process to reduce radiation damage during data collection. A detailed account of this procedure is currently available online at http://bit.ly/2f3pTez (*Do's and Don'ts of Cryoelectron Microscopy: A Primer on Sample Preparation and High Quality Data Collection for Macromolecular 3D Reconstruction*).

Single-particle tomographic reconstruction of capsid structure uses the same principles as the now familiar medical CT (computerized tomography) scan. Many two-dimensional slices are obtained from multiple particles randomly oriented relative to the electron beam. Images of particles in similar orientation are averaged and then combined to generate a three-dimensional reconstruction. Resolution is particularly good for icosahedral capsids due to their large particle size, their high symmetry that enables averaging of many particles, and the regular packing of their constituent proteins. In the most favorable situations, reconstructions at 4 Å (0.4 nm) resolution have been obtained.

The same methodological approach has been applied to phage-infected cells. This short video clip (http://bit.ly/1TyoZ6b)

---

[32] Such techniques improve the resolution, where resolution is the minimum distance between two points that can be distinguished as two separate points.

[33] cryo-electron microscopy: a variant of transmission electron microscopy in which samples are prepared by rapid freezing, often by plunging into liquid ethane, and then imaged without other chemical treatments, such as staining.

visualizes a cyanobacterium (*Synechococcus*) infected by cyano-phage[34] Syn5. (You must have a subscription or institutional access to the journal *Nature* online to view this video.) First shown is the section-by-section tomogram, which is followed by the three-dimensional structure of the cell generated from the series of two-dimensional tomographic slices. Since it is difficult to visualize either internal cellular structures or assembling phages within the cell, an annotated volume rendering is provided next. Here labels identify the cellular structures and assembling virions within the bacterium, as well as virions attached to the cell surface. Attached virions that have already delivered their chromosome appear as empty capsids. Credit: Dai, W, C Fu, D Raytcheva, J Flanagan, HA Khant, X Liu, RH Rochat, C Haase-Pettingell, J Piret, SJ Ludtke. 2013. Visualizing virus assembly intermediates inside marine cyanobacteria. Nature 502:707-710.

A caveat: when a capsid is described as icosahedral, this does not mean that it is a crisp icosahedron such as you might create by folding origami paper, with straight lines along every edge and 12 sharp points at the vertices. Rather it means that the capsid proteins are positioned according to the rules of icosahedral symmetry. The resultant capsid displays the two-fold, three-fold, and five-fold axes of rotation characteristic of an icosahedron. The same principle applies to the helical symmetry of some phage capsids and tails.

Virion dimensions are typically given in nanometers (nm) and cellular dimensions in microns (μm); 1 μm = 1,000 nm. Image resolution, as well as some molecular dimensions, are often reported in angstroms (Å); 1 nm = 10 Å.

---

[34] cyanophage: a phage that infects a member of the Cyanobacteria (a phylum of photosynthetic Bacteria).

In the Spotlight

Biped (*Acidianus* Two-tailed Virus)
    Family: *Bicaudaviridae*
    Host: *Acidianus convivator*
    Lifestyle: Non-lytic
    Chromosome: Circular
        dsDNA; 62,730 bp
    Genes: 72[35]
    Virion: Spindle-shaped capsid,
        119 nm × 243 nm; tailspan,
        ~744 nm

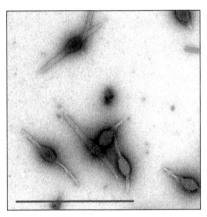

Figure 5: Biped's portrait. Negatively stained TEM of several Biped virions, some with not yet fully grown tails. Bar = 1 µm. Courtesy of David Prangishvili, Institut Pasteur, Paris, France.

Biped infects *Acidianus convivator*, an acid-loving, hyperthermophilic[36] crenarchaeon[37] that thrives in bubbling hot springs. Its spindle-shaped virion architecture is found only in phages that infect Archaea, notably those living in extreme environments. Unlike the other spindled-shaped phages, Biped has two-long tails that grow from the capsid after the virion has extruded from its host.

---

[35] I am defining *gene* here to mean a segment of the chromosome that is either predicted or known to encode a protein. Some phage chromosomes also encode non-translated RNAs such as tRNAs.

[36] hyperthermophile: an organism or virus that thrives at temperatures of 80° C and above.

[37] Crenarchaeota: a major phylum within the domain Archaea that includes many hyperthermophiles.

Chimera (Enterobacteria phage P22)
   Family: *Podoviridae* or *Siphoviridae*
   Host: *Salmonella typhimuriam*
   Lifestyle: Temperate
   Chromosome: Linear
      dsDNA; 41,724 bp
   Genes: 67
   Virion: Icosahedral capsid, 60
      nm diameter; short tail

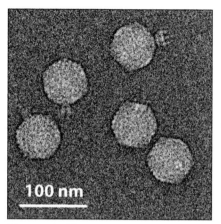

Like its mythological namesake,
Chimera is part one creature,
part another. Its virion morphology, including its Stubby-
like short tail, indicate its kin-
ship with the podoviruses; its
genome and temperate lifestyle

Figure 6: Chimera's portrait. Cryo-
EM image courtesy of Wah Chiu,
Baylor College of Medicine.

betray its close relationship to the siphoviruses, such as Lancelot and
Temperance. Chimera has proven to be very useful to researchers as
a tool to transfer genes between bacterial cells by generalized transduction.[38] The chromosome of its host (*Salmonella*) contains numerous
sequences that resemble, to varying degrees, Chimera's own packaging signal. As a result, about 2% of its virions contain approximately
42,000 bp of DNA from various regions of the host chromosome instead of a 41,724 bp phage chromosome. When such a virion "infects"
a related cell, the cell receives about 50 potentially useful genes from a
conspecific[39] instead of an invading phage.

---

[38] generalized transduction: the conveyance of cellular genes between hosts by phage
   virions due to DNA packaging errors.
[39] conspecific: a member of the same species.

Cowboy (*Salmonella* phage χ)
   Family: *Siphoviridae*
   Host: *E. coli, Salmonella*
   Lifestyle: Lytic
   Chromosome: Linear dsDNA;
      59,407 bp
   Genes: 75
   Virion: Icosahedral capsid,
      66 nm diameter; flexible,
      noncontractile tail, 220-230
      nm long; single tail fiber 200-
      220 nm long

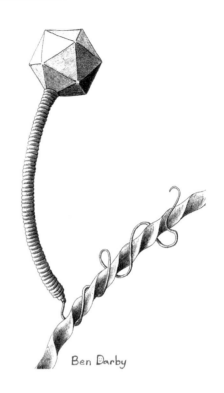

Figure 7: Cowboy's portrait. Original drawing by Ben Darby. Previously published in *Life in Our Phage World* by Rohwer, F, M Youle, H Maughan, N Hisakawa. 2014. Wholon. Used with permission.

Many motile Bacteria and Archaea swim briskly about powered by one or more rapidly rotating flagella.[40] Multiple flagella, all rotating in the same direction at more than one hundred rpm, propel *E. coli*, *Salmonella*, and their close kin. They are also targets for skillful flagellotropic phages such as Cowboy. Equipped with a single long tail fiber with a curled end, Cowboy lassos a rotating flagellum and rides it to the cell surface.

---

[40] flagellum (plural, flagella): an appendage of many prokaryote cells that is anchored in the cell membrane and rotates to propel the cell through the milieu.

Dynamo (*Bacillus* phage φ29)
   Family: *Podoviridae*
   Host: *Bacillus subtilis*
   Lifestyle: Lytic
   Chromosome: Linear
      dsDNA; 19,282 bp
   Genes: 27
   Virion: Prolate[41] icosahedral
      capsid, 45 × 54 nm; short,
      noncontractile tail, 38
      nm long; 12 appendages
      (tailspikes) attached to
      the neck; 55 head fibers,
      ~25 nm long

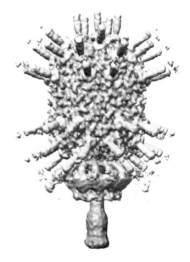

Figure 8: Dynamo's portrait. A cryo-EM single-particle reconstruction of Dynamo's virion at 30 Å resolution. The unusual head fibers function during attachment of the virion to the cell wall. Courtesy of EMDB (EMD-1506). Original publication: Xiang, Y, MC Morais, DN Cohen, VD Bowman, DL Anderson, MG Rossmann. 2008. Crystal and cryoEM structural studies of a cell wall degrading enzyme in the bacteriophage φ29 tail. Proc Natl Acad Sci 105:9552-9557.

Dynamo has a relatively short dsDNA chromosome, and it also has one of the smaller capsids. The net result is that it has the task of stuffing a 6.6 μm long chromosome into a prolate icosahedral capsid that is only 42 × 54 nm.[42] While other tailed phages face this same challenge and meet it in the same way, Dynamo's strong packaging motor has been investigated in greater detail. It is estimated to be eight to 25 times stronger than the myosin motor protein in our muscles and twice as strong as RNA polymerase (RNAP), itself a respected, strong molecular motor. Dynamo is also notable for the exceptional functionalities of its DNA polymerase that have made this enzyme the choice of researchers for DNA amplification.

---

[41] prolate: (adj.) elongated; for a sphere (or an icosahedron), being elongated such that the distance between the poles is greater than the diameter at the equator.

[42] A movie is available (http://bit.ly/2912GtJ) that shows a single particle reconstruction at 7.8 Å resolution of the virion of a fiberless Dynamo mutant. The initial rotation of the virion around both axes is followed by a series of longitudinal sections that show the arrangement of the packaged DNA within the capsid. Courtesy of Protein Data Bank Japan (PDBj). EM Data Bank (EMDB)/1420. Original publication: Tang, J, N Olson, PJ Jardine, S Grimes, DL Anderson, TS Baker. 2008. DNA poised for release in bacteriophage φ29. Structure 16:935-943.

Fickle (*Bordetella* phage BPP-1)
    Family: *Podoviridae*
    Host: *Bordetella bronchiseptica*
    Lifestyle: Lytic
    Chromosome: Linear dsDNA;
        42,493 bp
    Genes: 49
    Virion: Icosahedral capsid,
        ~70 nm diameter; short tail;
        six tailspikes; six tail fibers,
        each with a globular "foot"
        at the end

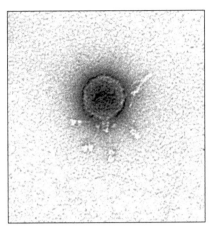

Figure 9: Fickle's portrait. TEM showing Fickle's short tail and long tail fibers with globular "feet". Courtesy of Mari Gingery, independent researcher.

Fickle earned its nickname for its ability to switch from infecting one host strain or species to another, a practice known as viral tropism switching. This it does out of necessity, not caprice, in order to counter the periodic disappearance of its usual receptor from the surface of its host. To make the switch, it introduces targeted point mutations that alter a few specific amino acids in the protein located at the tip of the tail fibers. This change suffices to enable Fickle to recognize and bind a different receptor. The same mechanism enables Fickle to reliably switch back to its original receptor.

Fusion (PM2)

Family: *Corticoviridae*
Host: *Pseudoalteromonas*
Lifestyle: Lytic
Chromosome: Circular
    dsDNA; 10,079 bp
Genes: 22
Virion: Icosahedral
    capsid, 60-63 nm
    diameter; internal lipid
    membrane

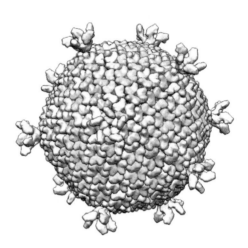

Figure 10: Fusion's portrait. A reconstruction of Fusion's capsid structure as determined by X-ray diffraction to 7.0 Å resolution. Courtesy of Protein Data Bank (2W0C). Original publication: Abrescia, NG, JM Grimes, HM Kivelä, R Assenberg, GC Sutton, SJ Butcher, JK Bamford, DH Bamford, DI Stuart. 2008. Insights into virus evolution and membrane biogenesis from the structure of the marine lipid-containing bacteriophage PM2. Mol Cell 31:749-761.

The *Corticoviridae* is the smallest of the phage families, Fusion being its only known member. Fusion is one of three phages featured in this book, all distinctly different, whose icosahedral capsids incorporate an internal or external lipid membrane (the other two being Shy and Slick). Each of the three uses its membrane in a different way to facilitate chromosome delivery. During infection, nearly 100% of all phages leave their intact, but now empty, capsid attached to the outer surface of the invaded cell. Fusion does not. Instead, upon adsorption its capsid falls apart to expose the internal membrane. This membrane fuses with the outer membrane of its soon-to-be host to deliver the phage chromosome into the cell.

Independence (Enterobacteria phage N15)

   Family: *Siphoviridae*
   Host: *E. coli*
   Lifestyle: Temperate
   Chromosome: Linear dsDNA; 46,375 bp
   Genes: 61
   Virion: Icosahedral capsid, ~50 nm
      diameter; flexible, noncontractile
      tail, ~141 nm long; small, brush-like
      structure at tail tip

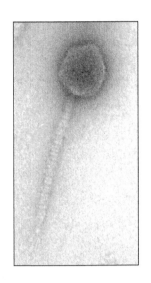

Figure 11: Independence's portrait. Negatively stained TEM courtesy of Nikolai V. Ravin, Institute of Bioengineering, Research Center of Biotechnology, Russian Academy of Sciences.

The vast majority of prophages integrate into the host chromosome, but not Independence's prophage. This one maintains itself as a separate, linear dsDNA molecule in the virocell cytoplasm. This strategy causes two specific problems. First, specialized maneuvers are required in order to replicate a linear molecule of dsDNA all the way to both ends. This predicament is resolved by eukaryotes, including ourselves, by using telomeres. Most phages and prokaryotes avoid this issue by circularizing their linear chromosome before replication. Independence, however, has its own solution, and replicates its chromosome from end to end each time the virocell divides. This, in turn, creates the second problem: how to ensure that one of the two daughter prophages goes to each daughter cell. Independence combines significant innovations and the theft of genes from skilled plasmids[43] to deliberately partition one prophage to each cell.

---

[43] plasmid: an independently replicating DNA molecule, separate from the cell's chromosome(s), that is found in many prokaryotes. Plasmids are inherited vertically, but also are sometimes transferred horizontally to another cell by conjugation.

Lancelot (Enterobacteria phage HK97)

    Family: *Siphoviridae*

    Host: *E. coli*

    Lifestyle: Temperate

    Chromosome: Linear
       dsDNA; 39,732 bp

    Genes: 62

    Virion: Icosahedral capsid,
       55 nm diameter; flexible,
       noncontractile tail, 177
       nm long

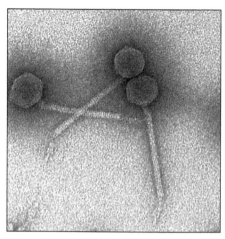

Figure 12: Lancelot's portrait. Negatively stained TEM courtesy of Robert Duda, University of Pittsburgh.

Like the legendary Knights of the Round Table, Lancelot's virions venture forth protected by chain mail armor. Whether fabricated by humans or phages, chain mail is composed of many small, interlocking rings that transform a minimum of material into a strong, flexible, and resilient coat. Our armor is a cumbersome, costly, add-on layer, but Lancelot's is an integral part of its capsid formed by the interlocking of the essential capsid proteins. This tactic enables its thin capsid to withstand the pressure generated by its packaged chromosome. Its chain mail actually saves on materials because related phages that lack it, such as Temperance, need to add glue proteins to strengthen their capsids.

Lander (Enterobacteria phage T4)

Family: *Myoviridae*

Host: *E. coli*

Lifestyle: Lytic

Chromosome: Linear dsDNA; 168,903 bp

Genes: 280

Virion: Prolate icosahedral capsid, 85 nm diameter, 115 nm long; contractile tail, 21 nm diameter, 100 nm long; long tail fibers, 145 nm long

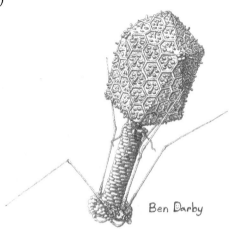

Figure 13: Lander's portrait. Original drawing by Ben Darby. Previously published in *Life in Our Phage World* by Rohwer, F, M Youle, H Maughan, N Hisakawa. 2014. Wholon. Used with permission.

Lander is the poster child of the phage world. Its archetypal virion is printed on T-shirts, reproduced on coffee mugs, sculpted into earrings. Each virion is assembled from more than 40 different proteins by a precise, stepwise, assembly sequence. It dedicates as many genes to virion construction as numerous phages have in their chromosome. Lander was one of the seven phages of *E. coli* selected in the mid-1940s as the "type" phages. With all researchers investigating the same phages, the field progressed rapidly. Lander became one of the two most-studied phages; thus, its numerous appearances in this book. Its virion is commonly depicted with all six tail fibers reaching downward like predatory claws, which also brings to mind images of a lunar lander preparing to touch down. In actuality, its virions travel with most fibers held close. More are lowered only when this Lander has landed on the surface of a potential host.

Minimalist (Enterobacteria phage Qβ)
   Family: *Leviviridae*
   Host: *E. coli*
   Lifestyle: Lytic
   Chromosome: Linear ssRNA;
      4,215 nt
   Genes: 3
   Virion: Icosahedral capsid,
      ~27 nm diameter

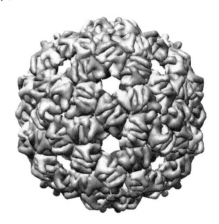

Coliphage Minimalist does more with less than most any other phage. It carries out an entire lytic infection cycle with a mere three genes. Moreover it does this even though it has to encode a protein not required by phages with DNA chromosomes: the replicase[44] that it needs to duplicate its RNA chromosome. Its capsid is simple, composed of the main capsid protein plus one copy of a multi-tasking protein that assists with adsorption and host lysis.

Figure 14: Minimalist's portrait. A reconstruction of Minimalist's capsid structure as determined by X-ray diffraction to 3.5 Å resolution. Its capsid is assembled from 180 copies of the major capsid protein plus one copy of the maturation protein (not shown). Courtesy of Protein Data Bank (1QBE). Primary publication: Golmohammadi, R, K Fridborg, M Bundule, K Valegard, L Liljas. 1996. The crystal structure of bacteriophage Qβ at 3.5 Å resolution. Structure 4: 543-554.

It adsorbs to the sides of the hollow sex pili that are formed on occasion by its hosts (e.g., *E. coli*) to facilitate DNA transfer from one cell to another.[45] In order to get by without any specialized regulatory proteins to coordinate its infection activities, Minimalist relies on the secondary structure[46] of its chromosome to govern gene activity (see "Production Management" on page 87).

---

[44] replicase: an RNA-dependent RNA polymerase (RdRP), i.e., an enzyme that synthesizes RNA using RNA as the template.

[45] conjugation: a form of bacterial "sex" in which DNA is transferred from one bacterium to another through cell-to-cell contact.

[46] secondary structure (nucleic acid): the three-dimensional structure of a DNA or RNA molecule that results from the formation of intermolecular or intramolecular hydrogen bonds (and other weak bonds). Examples: double helix, stem loop. (See "Nucleic Acid Structure" on page 56.)

Nerd (*Caulobacter* phage φCbK)
  Family: *Myoviridae*
  Host: *Caulobacter crescentus*
  Lifestyle: Lytic
  Chromosome: Linear dsDNA; 215,710 bp
  Genes: 338
  Virion: Prolate icosahedral capsid, 56 nm
      diameter, 205 nm long; noncontractile tail,
      21 nm diameter, 290 nm long; tail fibers,
      145 nm long; head filament, ~200 nm long

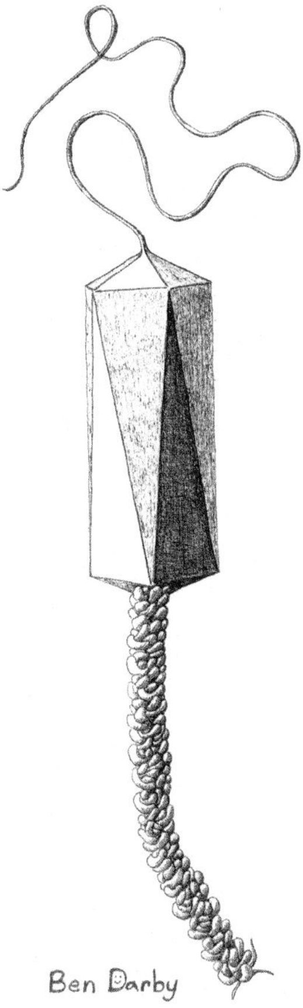

Figure 15: Nerd's portrait. Original drawing by Ben Darby. Previously published in *Life in Our Phage World* by Rohwer, F, M Youle, H Maughan, N Hisakawa. 2014. Wholon. Used with permission.

Nerd's bacterial host, *Caulobacter crescentus*, is an atypical bacterium in that it has a distinct life cycle. When first spawned, these cells swim about propelled by a flagellum. Later, they settle down, attach to a surface, and bud new swimmer cells. Nerd preys only on the swimmers, using their flagella as its landing platform. While there are numerous other flagellotropic phages,[47] they all (with the exception of a close relative of Nerd) use their tail fibers to capture a flagellum (see "Cowboy" on page 23). Nerd's immediate family is, so far, unique in using their heads instead, specifically using the long filament attached directly to the apex of their capsid. How Nerd assembles these filaments and ensures that they are attached at the correct vertex remains a puzzle.

_____

[47] flagellotropic phage: a phage whose virion recognizes and adsorbs to a flagellum of its potential host.

Pharaoh (*Sulfolobus* turreted icosahedral virus STIV)

Family: Unassigned (see
   Pina et al., 2011, in
   "Further Reading" on
   page 59)
Host: *Sulfolobus
   solfataricus*
Lifestyle: Lytic
Chromosome: Circular
   dsDNA; 17,663 bp
Genes: 36
Virion: Icosahedral
   capsid, 72 nm
   diameter; internal
   lipid membrane;
   turret-like structures
   at the vertices

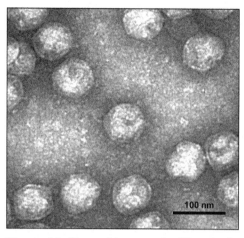

100 nm

Figure 16: Pharaoh's portrait. Negatively stained TEM courtesy of Sue K. Brumfield and Mark Young, both at Montana State University.

Not so long ago, it seemed obvious that a bubbling mud pot, geyser, or hot spring was inimical to all life. What could live at temperatures above 80° C and at a pH <3? The answer is Crenarchaeota and their diverse phages (including Biped and Pharaoh). Such extreme environments have delighted phage hunters with numerous surprises. The inventive phages that infect one archaeal genus (*Sulfolobus*) include some with filamentous, rod-like, spindle-shaped, and droplet-shaped virions – and also Pharaoh with its unusual turreted icosahedral capsid. Pharaoh is famous for the towering proteinaceous pyramids it constructs on the surface of its host; an infection ends with the pyramids opening to release the waiting virions. The structure of Pharaoh's major capsid protein reveals its relatedness to some phages that infect Bacteria and Eukarya. This finding suggests that far back in Pharaoh's family tree was an ancestor that infected the early cellular life forms that later gave rise to all cellular life that we see today.

Positivist (bacteriophage SPP1)
   Family: *Siphoviridae*
   Host: *Bacillus subtilis*
   Lifestyle: Lytic
   Chromosome: Linear
      dsDNA; 44,010 bp
   Genes: 280
   Virion: Icosahedral capsid,
      ~60 nm diameter;
      noncontractile tail, 177
      nm long plus tail fibers

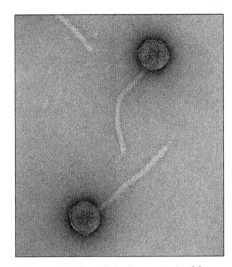

Figure 17: Positivist's portrait. Negatively stained TEM courtesy of Rudi Lurz, Max-Planck Institut for Molecular Genetics, Berlin, Germany and Paulo Tavares, Department of Virology, Gif-sur-Yvette, France.

Most of the phages featured in this book infect enteric Gram-negative Bacteria, most often *E. coli* or *Salmonella*. This proportion is misleading because phages with Gram-positive hosts are just as abundant, even though they have been less studied. Positivist is one of those phages that infects a Gram-positive bacterial species, *Bacillus subtilis*. The Gram-positive cell surface presents different challenges for a phage. The outermost layer is a thick cell wall that blocks phage access to its receptor and point of entry on the cell membrane. Positivist overcomes this with its two-step approach.

Shy (*Pseudomonas* phage φ6)
    Family: *Cystoviridae*
    Host: *Pseudomonas savastanoi*
        pv. phaseolicola
    Lifestyle: Lytic
    Chromosome: Linear dsRNA,
        segmented
        Segment L: 6,374 bp
        Segment M: 4,063 bp
        Segment S: 2,948 bp
    Genes: Segment L: 4 genes
        Segment M: 4 genes
        Segment S: 4 genes
    Virion: Icosahedral capsid, 85
        nm diameter; external lipid
        membrane

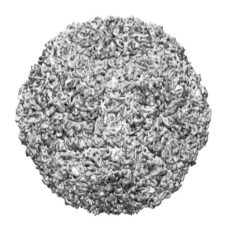

Figure 18: Shy's portrait. A reconstruction of Shy's virion structure as determined by cryo-EM to 7.5 Å resolution. Courtesy of Protein Data Bank (4BTQ). Primary publication: Nemecek, D, E Boura, W Wu, N Cheng, P Plevka, J Qiao, L Mindich, JB Heymann, JH Hurley, AC Steven. 2013. Subunit folds and maturation pathway of a dsRNA virus capsid. Structure 21:1374-1383.

Cells make many noncoding[48] dsRNA molecules, but these are short molecules – typically no more than about 25 bp. A dsRNA chromosome delivered by a phage into any cell would be immediately recognized as foreign and set upon by cellular RNases. Thus, shyness is a matter of necessity for a phage such as Shy that uses dsRNA for its chromosome. To avoid death by RNase, it protects its chromosomes by keeping them within a protein capsid throughout the entire lytic replication cycle, from entry to exit. Although this tactic is seen in some eukaryotic viruses, among the phages it is known in only this one small family. Its implementation required the evolution of different modes of host entry and chromosome replication, as well as changes to virion architecture and assembly. As if that weren't enough of a challenge for one phage, Shy also has a segmented genome. To assemble infectious progeny, Shy must package one, and only one, copy of each of its three chromosomes inside each virion. This it does with a high rate of success.

---

[48] noncoding RNA: an RNA molecule that is not an mRNA, i.e., whose sequence is not translated into the amino acid sequence of a protein. Some noncoding RNAs function in translation or regulate cellular activities.

Skinny (Enterobacteria phage Ff[49])

Family: *Inoviridae*

Host: *E. coli*

Lifestyle: Non-lytic

Chromosome: Circular ssDNA; 6,407 nt

Genes: 10

Virion: Helical filament, 7 nm diameter, 700-950 nm long

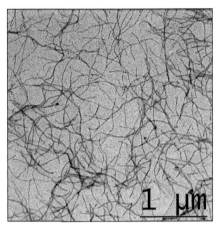

Figure 19: Skinny's portrait. TEM of many Skinny virions courtesy of Marcus Drechsler, University of Bayreuth, Germany.

By their very existence, filamentous phages demonstrate that a phage can replicate for countless generations and produce copious progeny without ever once lysing a virocell. However, the small number of phages that use this tactic bespeaks the inherent limitations. Skinny's virions are remarkably long for a phage, but even so their length limits these phages to small genomes with fewer than a dozen genes. Entry and exit procedures also pose unique problems for Skinny. However, like all phages, it knows its host intimately and recruits numerous host capabilities to assist. It successfully competes for its share of the potential *E. coli* hosts with phages such as Lander with its 280 genes. Is Skinny less exploitative than those phages that lyse their virocell? Is this collaboration or enslavement?

---

[49] enterobacteria phage Ff: a group of filamentous phages that includes f1, fd, M13, and others, all of whose genome sequences share 98% identity. They all infect *E. coli* through interaction with the F pilus.

Slick (Enterobacteria phage PRD1)
   Family: *Tectiviridae*[50]
   Host: *E. coli* harboring a
      conjugative plasmid
   Lifestyle: Lytic
   Chromosome: Linear
      dsDNA; 14,927 bp
   Genes: 31
   Virion: Icosahedral capsid,
      70 nm diameter; internal
      lipid membrane; spikes at
      the vertices

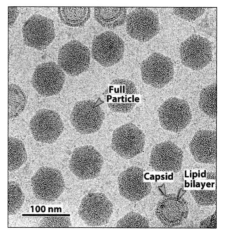

Slick is one of the phages that encloses its entire chromosome inside a membrane sac, then surrounds both membrane and chromosome with a protein shell. To construct its membrane, Slick steals lipids from the virocell and then embeds numerous proteins of its own within it. Upon adsorption to a

Figure 20: Slick's portrait. Cryo-EM image of Slick virions, including a few unfilled capsids in which the internal lipid membrane is visible. The spiral pattern visible in some of the filled capsids results from the combination of the capsid protein arrangement and the DNA spooled inside. Courtesy of Nicola G. A. Abrescia, CIC bioGUNE, Spain, and Dennis H. Bamford, University of Helsinki, Finland.

host, these proteins revamp the sac, turning it into a lipid chute for passage of the phage chromosome en route to the cytoplasm. So far, we know of only one small phage family, the *Tectiviridae*, that uses this slick trick. Slick's classification has been slippery, too. Although its virion morphology places Slick in the *Tectiviridae*, genome-based classification would assign it instead to the *Podoviridae*. It is not a typical podovirus, either, as it is a member of a small subfamily that uses an atypical, protein-primed mechanism of DNA replication.

---

[50] PRD1 is classified as a tectivirus based on its virion morphology, but analysis of its genome sequence assigns it to the *Podoviridae*.

Spindly (Lemon-shaped archaeal virus His1)
  Family: unassigned; floating genus
    Salterprovirus
  Host: *Haloarcula hispanica.*
  Lifestyle: Non-lytic
  Chromosome: Linear dsDNA; 14,462 bp
  Genes: 35
  Virion: Spindle-shaped capsid, ~40 nm
    diameter, ~92 nm long; tail, 12 nm long

Figure 21: Spindly's portrait. Cryo-EM reconstruction
of Spindly's spindle-shaped (lemon-shaped) capsid
at 20 Å resolution. Courtesy of Protein Data Bank in
Europe (EMD-6222). Primary publication: Hong, C,
MK Pietilä, CJ Fu, MF Schmid, DH Bamford, W Chiu.
2015. Lemon-shaped halo archaeal virus His1 with
uniform tail but variable capsid structure. Proc Natl Acad Sci 112:2449-2454. A
splendid movie prepared from virion cryo-EM data (http://bit.ly/291wW7f) reveals the tail structure, including the six tailspikes. Source: see above.

Spindle-shaped virions are found only among phages that infect Archaea. In extremophilic environments where Archaea outnumber
Bacteria, such as acidic hot springs and solar salterns, spindles are the
most abundant morphotype. Among these phages are some that assemble spindles with a short tail, some with a long tail, and a few who
grow two long tails. Based on their spindle morphology, all of these
phages had been assigned to a single viral family, the *Fuselloviridae*,
all of whose known members infect hyperthermophiles. Then along
came Spindly. Its morphology–spindle-shaped virion with one short
tail–suggests that it is a fusellovirus. Likewise, its capsid protein is
similar to that of the other fuselloviruses, which suggests they share a
common ancestor. However, it differs from the others in that its host
is not a hyperthermophile, but instead is an extremely halophilic eu-
ryarchaeon.[51] Moreover, its chromosome is linear dsDNA rather than
circular, which in turn necessitates a different mode of chromosome
replication. These differences leave Spindly currently an orphan without a family assignment.

---

[51] Euryarchaeota: a very diverse phylum within the domain Archaea that includes
anaerobic methane producers (methanogens) and extreme halophiles that require a
high salt concentration environment, as well as some thermophiles.

Stubby (Enterobacteria phage T7)
Family: *Podoviridae*
Host: *E. coli*
Lifestyle: Lytic
Chromosome: Linear dsDNA;
 39,937 bp
Genes: 60
Virion: Icosahedral capsid,
 60–61 nm diameter; tail, 23
 nm long

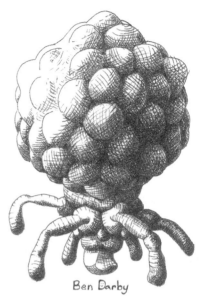

Ben Darby

In the 1940s, the researchers known as the Phage Group saw the advantages of having everyone in this new and burgeoning field focus on the same few phage types. To that end, they selected seven type phages isolated from sewage or human feces: three myoviruses including Lander, two podoviruses including Stubby, and two siphoviruses. Lander is the most extensively studied of these T-phages, but Lander's strategies are not universal, nor are they even shared among all the T-phages. Time and again, Stubby has provided a distinct counterpoint to Lander's story. Its strategies require only one-fifth as many genes. While Lander indulges in an extravagant tail, Stubby demonstrates that a short tail, assembled on the fly, can accomplish the same tasks. It also has its own distinctive mechanisms for both chromosome delivery and chromosome protection on arrival. Stubby is definitely a T-phage worth studying.

Figure 22: Stubby's portrait. Original drawing by Ben Darby. Previously published in *Life in Our Phage World* by Rohwer, F, M Youle, H Maughan, N Hisakawa. 2014. Wholon. Used with permission.

Temperance (Enterobacteria phage λ)
  Family: *Siphoviridae*
  Host: *E. coli*
  Lifestyle: Temperate
  Chromosome: Linear dsDNA;
    48,502 bp
  Genes: 74
  Virion: Icosahedral capsid, 50 nm
    diameter; flexible, noncontractile
    tail, 150 nm long; 0, 4, or 6 tail
    fibers

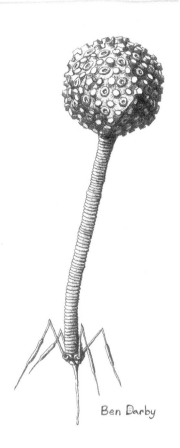

Figure 23: Temperance's portrait. Original drawing by Ben Darby. Previously published in *Life in Our Phage World* by Rohwer, F, M Youle, H Maughan, N Hisakawa. 2014. Wholon. Used with permission. A TEM made by Bob Duda, University of Pittsburgh, is available at http://bit.ly/2dP8iuV.

Ben Darby

Temperance has been the most intensely studied phage. As the field of molecular biology blossomed, Temperance was there as the model system of choice. Investigation of its replication tactics provided the foundation for our current understanding of many cellular processes, including mechanisms of gene regulation, both homologous and site-specific recombination, and the role of cellular chaperones.[52] Temperance's method of host lysis reigned as the paradigm for the field, the model to which other mechanisms have long been compared. Being a temperate phage, when its chromosome arrives in a host cell, it chooses, based on current conditions, to initiate immediate replication leading to lysis or to instead reside long-term as a prophage within the chromosome of the virocell. The "genetic switch" that governs this decision has been dissected in great detail.

---

[52] chaperone: a protein that assists the correct folding of other protein molecules or recognizes misfolded proteins and helps them to refold correctly.

Thief (Enterobacteria phage P4)
  Family: *Podoviridae*
  Host: *E. coli*
  Lifestyle: Temperate
  Chromosome: Linear dsDNA;
    11,623 bp
  Genes: 14
  Virion: Icosahedral capsid,
    45 nm diameter; flexible
    noncontractile tail, 135 nm
    long; 6 tail fibers, 45-50 nm
    long

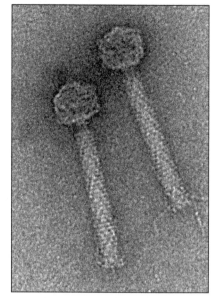

Thief steals proteins. Specifically, it steals the structural proteins[53] needed to assemble its virions from a "helper" phage. When classified by virion architecture, Thief is classified as a myovi-

Figure 24: Thief's portrait. TEM courtesy of Terje Dokland, University of Alabama at Birmingham.

rus, like its helper phage P2. However, genomic analysis revealed its podovirus nature. This protein theft is not random highway robbery, but the result of a highly evolved, intimate relationship between Thief and helper. Thief intercepts and falsifies the regulatory messages that direct helper activities. Moreover, Thief has adjusted its temperate lifestyle to increase the probability that a helper will be on hand and producing proteins when Thief needs them. Most of the information included in this book was obtained from research on single phage-host pairs. As the interactions between Thief and its helper demonstrate, the complex phage communities in natural ecosystems offer innumerable other possibilities for both cooperation and exploitation.

---

[53] structural protein: a protein that is a structural component of the **capsid** (or tail or membrane) of a mature virion.

Yoda (coliphage φX174)

 Family: *Microviridae*

 Host: *E. coli*

 Lifestyle: Lytic

 Chromosome: Circular
   ssDNA; 5,386 nt

 Genes: 11

 Virion: Icosahedral
   capsid, ~27 nm
   diameter; spikes at the
   vertices

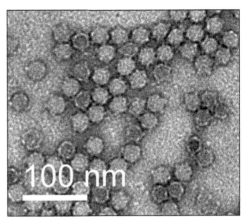

Figure 25: Yoda's portrait. TEM courtesy of Yingyuan Sun, Purdue University; Michael Rossmann, Purdue University; and Bentley Fane, University of Arizona.

Yoda's small chromosome was sequenced in 1977 – the first complete DNA-based genome to be sequenced. It had been puzzling how Yoda could possibly encode its eleven proteins in a genome of only 5,386 nucleotides. The sum of the gene lengths required was greater than the length of its chromosome. That calculation assumed that the eleven genes were all lined up in single file along the chromosome – a reasonable assumption at the time because that had been the case for all previously characterized genes. The idea that genes could overlap was not only counterintuitive, but it even seemed bizarre. Commonsense assumptions notwithstanding, when Yoda's genome was sequenced, three genes were found to overlap others (see Figure 26). This was so unexpected that some people seriously considered the possibility that Yoda's genome encoded a message from an advanced civilization elsewhere in the universe.[54] Having now sequenced more than 5,000 viral genomes, we know that many viruses use this trick to do more with less. Yoda also uses a trick to make do with a smaller capsid. Its ssDNA chromosome conveys the same amount of information as would its dsDNA counterpart, but requires less virion cargo space.

---

[54] See Yokoo and Oshima 1979 in "Further Reading" on page 59.

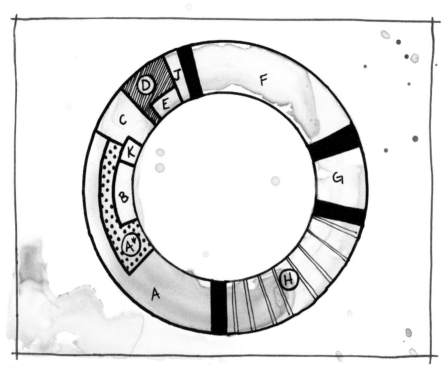

Figure 26: Yoda's circular, ssDNA genome. Note the overlapping genes in different reading frames.

Essentials of Molecular Biology

Not for You?

In subsequent chapters, I will assume you know the molecular biology basics presented below. If you are already familiar with these topics, you may want to skim, or entirely skip, this section. If not, digesting this brief overview will enable you to appreciate the phage "thoughts" that follow.

DNA makes RNA makes protein. This often quoted mantra is the central dogma of molecular biology. It summarizes in five words how the genetic information encoded in DNA is first transcribed into RNA and then translated into the amino acid sequences of the multitude of specific proteins needed by each cell. The phages have mastered – perhaps in some cases invented – the basics of molecular biology. The replication of genetic information and the synthesis of proteins are the essence of the phage life. Phages have also repeatedly pushed the envelope by evolving unique variations on the common practices, variations that you won't find anywhere in cellular life.

Proteins

Why is so much cellular and phage effort devoted to the reliable synthesis of proteins? Because these macromolecules comprise the great majority of their structural components and because proteins are the catalysts for many chemical reactions carried out by cells, including the synthesis of the very DNA and RNA used to fashion them. Proteins are vital to every phage activity. They form the structure of virions, catalyze the replication of phage chromosomes, carry out virocell lysis, and serve as the regulators that orchestrate efficient phage replication. A protein molecule is a linear chain of amino acids covalently linked head-to-tail by peptide bonds.[55] They are all built from the same 20 or

---

[55] peptide bond: the covalent bond linking consecutive amino acids in a protein chain. The acid portion of one amino acid (a carboxyl group) is joined to the amino portion (the amino group) of the next amino acid.

so common amino acids, along with a few minor ones, arranged in the particular sequence that defines that particular protein. Much cellular energy is invested in these peptide bonds, and much machinery and care is devoted to ensuring the amino acid sequence is correct in copy after copy, generation after generation.

A single protein molecule, containing hundreds or even thousands of amino acids, may be composed of two or more regions, or domains.[56] Each domain may catalyze a step in a multi-step reaction, determine the cellular location of the protein, or bind the protein to a specific molecule. Protein function evolves rapidly when variants of a particular gene that are present in different strains or species have the opportunity to recombine.[57] New capabilities can result when domains are recombined in new ways. A protein molecule may function individually, or it may associate with other copies of the same protein to constitute the active homodimer,[58] homotrimer, etc. Similarly, different proteins can bind together as a heterodimer,[59] etc., to create a structural element or catalyze a reaction.

The correct amino acid sequence is necessary, but not sufficient, to ensure a functional protein. These large molecules must be properly folded, often into a specific, complex, three-dimensional configuration (see "Protein Structure" on page 57). All cells contain a group of proteins, the chaperones, that assist the folding of newly made proteins, as well as the correct refolding of misfolded ones. Most phages rely on the chaperones of their virocell to fold their proteins, but a few also encode their own.

Proteins that act primarily as enzymes are often assigned a name with the suffix –ase. The rest of the name typically indicates the reaction

---

[56] protein domain: a unit of protein structure that can evolve and function independent of the other domains in that protein. A multi-domain protein may carry out a coordinated action in which different domains are responsible for distinct functions such as the localization of the protein in the cell, substrate binding, ATP hydrolysis, and catalysis.

[57] recombination: in prokaryotes and phages, the integration of a segment of exogenous DNA (or RNA) into a DNA (or RNA) chromosome by the enzymatic cutting and rejoining of the DNA (or RNA) molecules.

[58] homodimer: a functional unit composed of two copies of the same macromolecule, most often a protein, joined by non-covalent bonds.

[59] heterodimer: a functional unit composed to two different macromolecules, most often proteins, joined by non-covalent bonds.

catalyzed or the name of a reaction substrate or product. For example, DNA polymerase catalyzes the synthesis of DNA by the polymerization of nucleotides, DNase catalyzes its cleavage, and DNA methylase adds methyl groups to its bases (see "Chromosomes" on page 45).

Chromosomes

The chromosomes in all cellular life forms are composed of double-stranded DNA (dsDNA) molecules, the famous double helix. Each strand of DNA is a linear polymer of building blocks, the nucleotides; each nucleotide includes three components – a five-carbon sugar, a phosphate group, and a ring-shaped carbon-containing moiety referred to as a base. A string of alternating sugars and phosphates forms the backbone of the nucleic acid, while a base attached to each sugar projects off to the side. The phosphate groups in the backbone are negatively-charged. Their mutual repulsion resists phage efforts to compact their DNA to a high density inside a virion (see "Packaging DNA" on page 124).

Inherited genetic information is encoded in the sequence of bases along the DNA backbone. The four different bases used comprise two complementary pairs (**G**uanine paired with **C**ytosine, and **A**denine paired with **T**hymine). Collectively, the phages employ a larger variety of bases in their DNA. Some variants are modifications of those four that have no impact on base pairing, but dserve to protect the DNA from attack by host or phage nucleases (see "Death by Nuclease" on page 64). The two strands in a dsDNA chromosome have complementary base sequences (see Figure 27). Each base in one strand binds weakly to its complement in the other strand via two or three hydrogen bonds.[60] Implicit in this structure is the mechanism for the accurate replication of information that underlies genetic inheritance. During replication, the two strands separate and each serves as a template for the synthesis of its complementary strand by DNA polymerase.[61] In general, prokaryotes and phages replicate a single circular chromosome, whereas

---

[60] Hydrogen bonds are weak, electrostatic interactions between a slightly positively-charged hydrogen atom and a slightly negatively-charged atom nearby. They are much weaker than covalent bonds in which shared electrons bind atoms together to form molecules.

[61] This mode of replication is termed semi-conservative because each daughter chromosome contains one old and one new strand.

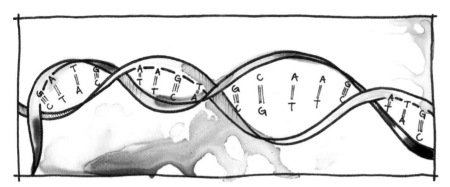

Figure 27: Double helix. The two strands in a dsDNA helix are conjoined at every nucleotide by non-covalent bonds (hydrogen bonds) between complementary bases: adenine to thymine (A-T), guanine to cytosine (G-C). These bonds are broken and the strands separated locally to provide the single-stranded template needed during replication and transcription.

eukaryotes replicate and segregate multiple linear chromosomes. Circular chromosome replication required solving a topological problem, that being that the daughter circles are intertwined and must be freed to go their individual ways. Complete end-to-end replication of a linear chromosome posed a different problem in that DNA polymerase can't place the first nucleotide unless it has a previous nucleotide, or surrogate, to which it can be linked. The eukaryote workaround is to cap our chromosome ends with repetitive sequences (telomeres) and provide an enzyme to replace lost nucleotides. Eccentric phages, such as Independence, who replicate a linear dsDNA molecule have a different solution (see "Independent Prophages" on page 248).

Although many phages package chromosomes of dsDNA inside their virions, a few transport only one strand (ssDNA) and then let cellular enzymes convert it into dsDNA after its arrival in the host cell. Others use chromosomes of RNA, either single-stranded or double-stranded. RNA differs from DNA only in that it contains a slightly different sugar (ribose, rather than deoxyribose) and the base thymine is replaced by uracil (U) that has the same hydrogen-bonding capabilities. Any phage that deviates from the cellular norm by using RNA must supply its own chromosome replication machinery, whereas those that use DNA can, if they so choose, rely on their host's enzymes. Cells are not equipped to replicate RNAs, but rather to synthesize them using DNA templates. RNA chromosomes pose other challenges to phage survival and to efficient phage protein synthesis (see "The Plight of RNA Chromosomes" on page 83 and "The Ingenious Minimalist" on page 92).

RNAs are versatile molecules that store genetic information, support several steps in protein synthesis, catalyze biochemical reactions (ribozymes[62]), and regulate many activities including gene expression and translation (riboswitches,[63] sRNAs, etc.). A well-respected hypothesis envisions that earliest life did not require either DNA or proteins. Instead, in this very early RNA World, RNA performed the essential functions of information storage, self-replication, and catalysis. Only later was the archiving of genetic information handed off to the more chemically stable DNA and catalysis to the more adept proteins. Today RNA continues to perform some of these other functions on a small scale, but it no longer constitutes the chromosomes of any cells. That some phages and other viruses do have RNA chromosomes suggests that viruses – whether they themselves are alive or not – have accompanied life from its earliest days.

A gene, as defined in this book, is a section of a chromosome whose base sequence encodes the amino acid sequence of one protein. Every three nucleotides in a gene specifies the insertion of one particular amino acid at the corresponding position in the resultant protein. Therefore, the genetic code is termed a triplet code. Each base in a gene can be thought of as a "letter" in a string of three-letter "words" or codons. How much information can be specified by such a simple code? Since there are four different bases, there are four possibilities for each of the three letters. The number of different triplets possible is thus $4 \times 4 \times 4$ or 64. In comparison, a two-letter code would offer only 4×4 or 16 possible words. Since there are about 20 commonly used amino acids, a two-letter code would be inadequate. A triplet code provides enough codons to not only designate all the common amino acids (and a few others), but also to signal "start translation here" and "stop translation here", as well as allow for some redundancy in the code.[64] This genetic

---

[62] ribozyme: an RNA molecule that functions as an enzyme, i.e., a catalytic RNA.

[63] riboswitch: an RNA molecule that regulates its own synthesis by binding to a specific metabolite or other small molecule.

[64] In this context, redundancy means that the same amino acid is specified by more than one codon. This has important ramifications, one of them being that many mutations that change a single base do not alter the amino acid sequence of the corresponding protein.

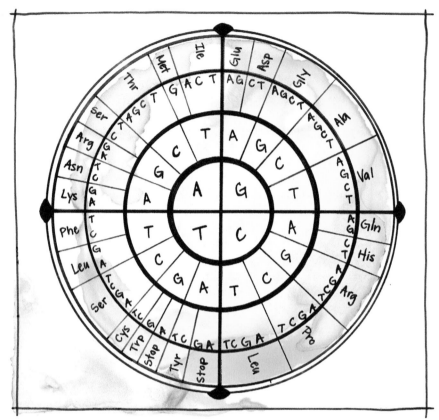

Figure 28: The DNA genetic code for 20 common amino acids plus two stop codons. **Ala**: Alanine; **Asn**: Asparagine; **Asp**: Aspartic Acid; **Arg**: Arginine; **Cys**: Cysteine; **Gln**: Glutamine; **Glu**: Glutamic Acid; **Gly**: Glycine; **His**: Histidine; **Ile**: Isoleucine; **Leu**: Leucine; **Lys**: Lysine; **Met**: Methionine; **Phe**: Phenylalanine; **Pro**: Proline; **Ser**: Serine; **Thr**: Threonine; **Trp**: Tryptophan; **Tyr**: Tyrosine; **Val**: Valine.

code (see Figure 28) is universal, used by all cells and viruses with only a few very minor tweaks – the exceptions that prove the rule.

Transcription

A DNA (or RNA) phage chromosome functions primarily as a transportable archive that can be delivered into a cell where it is read, as well as replicated. In eukaryote cells, the cellular enzymes that carry out these processes are sequestered along with the DNA within a membrane-bounded nucleus. By contrast, in prokaryotes the compacted cellular DNA forms a dense region in the cytoplasm, the nucleoid. As a result, the enzymes and other factors associated with the cell's chromosome that the phage needs to commandeer are accessible. The first step in the synthesis of a protein is the transcription of its gene, i.e., the synthesis of a complementary ssRNA molecule termed messenger RNA (mRNA). In each gene only one strand of the DNA is transcribed, the template strand. Its complementary strand, called the coding strand, encodes the same information as the transcribed mRNA (with a U substituted in the mRNA for each T in the DNA). That mRNA is termed positive-sense,[65] with its non-functional complement being negative-sense. In Bacteria and phages, a region of DNA that includes more than one contiguous gene – an operon – may be transcribed into one longer polycistronic[66] mRNA. This tactic is widely used to coordinate the expression of a group of genes whose protein products function together and are needed at the same time.

DNA transcription is the work of the RNA polymerase[67] holoenzyme[68] (RNAP), a complex of five proteins.[69] An additional protein, called a sigma factor, associates with the RNAP long enough to initiate transcription. RNAP works fast, adding up to 90 nucleotides per second to the growing transcript. A typical gene composed of 1000 nucleotides and encoding a protein of approximately 300 amino acids can be transcribed in as little as 11 seconds. And because RNAP corrects some of its mistakes, bacterial transcripts have only one "typo" in 10^5 nucleotides, thus only one per 100 transcripts of a typical gene. RNAP is also forceful. To advance along the DNA, it must continually break the bonds holding the double helix together ahead of it. This creates a "bubble" of single-stranded DNA that can serve as a template. Furthermore, whether this large enzyme complex moves along the DNA or ratchets the DNA past itself, it must overcome the significant drag exerted on these large molecules by the viscous environment. RNAP is

---

[65] positive-sense: refers to an RNA or DNA molecule whose sequence encodes the amino acid sequence of a protein.

[66] "Cistron" is another term for a gene that encodes a protein, i.e., encodes one chain of amino acids.

[67] RNA polymerase (RNAP): a holoenzyme that selectively transcribes regions of a dsDNA chromosome using the base sequence of one DNA strand as its template. It is also known as DNA-dependent RNA polymerase.

[68] holoenzyme: a functional enzyme complex composed of multiple components, e.g., the DNA polymerase and RNA polymerase holoenzymes.

[69] This discussion of transcription and its regulation applies specifically to Bacteria. The mechanism in Archaea is distinctly different, and bears some resemblance to that of the Eukarya.

one of the strongest motor proteins known, stronger than the myosin in our muscles.

Heretical phages, as well as other viruses and some cells, violate the central dogma of molecular biology and transcribe some information from RNA into DNA. These all make a reverse transcriptase,[70] an enzyme that uses an RNA template to synthesize complementary DNA. This step is central to the mechanism used by Fickle (see "Fickle" on page 210) to specifically alter its host range.

Transcriptional Regulation

Transcription offers opportunities for regulation of gene activity. Specific genes can be turned on and off, or their rate of transcription modulated. Upstream of the protein-coding segment are short regions that influence the rate of transcription of downstream genes. In Bacteria, one such region, the promoter, is recognized by the sigma factor associated with RNAP. RNAP initiates transcription nearby and later releases the sigma factor as it transcribes its way further along the DNA. Bacteria can make numerous sigma factors. *E. coli*, for example, has a repertoire of seven, each of which is synthesized only under particular growth conditions. Each one recognizes a different promoter sequence and turns on a different set of operons. Such a set of operons, which can be scattered throughout the chromosome, makes up a regulon. Bacteria make repressor proteins that bind to a specific promoter sequence to block RNAP access, as well as activator proteins that enhance RNAP-promoter interactions.

What do these mechanisms of gene regulation mean for a phage? The synthesis of phage proteins by the virocell requires, as the first step, transcription of the corresponding phage genes. This transcription, in turn, requires a match between the promoters in the phage chromosome and an available sigma factor. Phages routinely handle this need by encoding promoter sequences that match a sigma factor used by the host, but some phages see possibilities for greater exploitation here. They synthesize a sigma factor of their own that recognizes their proprietary gene promoters but not those of their host. This strategy not

---

[70] reverse transcriptase: an RNA-dependent DNA polymerase, i.e., an enzyme that synthesizes complementary DNA using single-stranded RNA as the template.

only enables synthesis of phage proteins, but also diverts the RNAPs associated with the phage's sigma factor away from host genes, thereby shutting down transcription of host genes. Disabled in this way, a newly infected host cell can't launch a counterattack.

These tactics for regulating transcription are not an option for phages with RNA chromosomes. Here all mRNAs are synthesized as chromosome-length ssRNAs using one strand of a dsRNA chromosome as the template. Every mRNA is not only polycistronic, but it also encodes every phage gene. In order to synthesize more of one protein than another, the phage must intervene during the next step – translation (see "The Ingenious Minimalist" on page 92).

Translation

The nucleotide sequence of an mRNA molecule is translated into the amino acid sequence of a protein molecule by the sophisticated protein-RNA structures known as ribosomes.[71] Each ribosome contains three different ribosomal RNAs plus more than 50 different ribosomal proteins. Cells also synthesize a collection of small RNAs known as transfer RNAs (tRNAs), one for each codon used. One arm of a tRNA displays an anticodon while another carries the specific amino acid specified by the complementary codon (see Figure 29). The enzymes[72] that attach the amino acid to the corresponding tRNA contribute significantly to the accuracy of translation. They reliably recognize the appropriate amino acid and tRNA, and also self-correct any incipient errors. As the ribosome translates the mRNA, a corresponding tRNA is matched to each codon and remains in the ribosome complex long enough for its attached amino acid to be transferred to the growing protein chain.

Translation begins with the attachment of a ribosome to an mRNA at a specific ribosome binding site near a start codon. Then, as the ribosome moves down the mRNA, one by one each codon is read and another amino acid is added to the lengthening protein. The addition of each amino acid uses the energy from one ATP to attach the amino acid to

---

[71] An animation illustrating protein translation is available online at http://bit.ly/2jG3LJ1.

[72] aminoacyl tRNA synthetases.

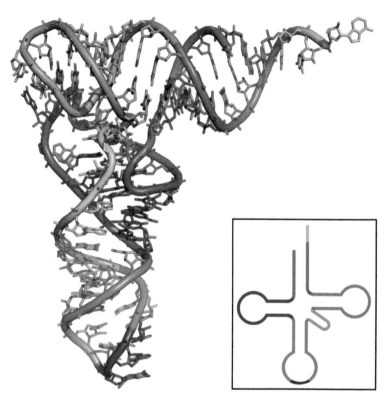

Figure 29: A typical tRNA cloverleaf structure. This example portrays a phenyl-alanine tRNA from yeast. The three-dimensional structure obtained by X-ray crystallography defines the position of each base as it projects from the back-bone. The amino acid acceptor stem (violet) terminates with the acceptor site (yellow). The anticodon stem-loop (blue) presents the anticodon (gray) to the mRNA. Courtesy of Yikrazuul. http://bit.ly/2dTpM9E

the tRNA plus the energy from two molecules of GTP to form the pep-tide bond and move the ribosome along the mRNA. The attachment of additional ribosomes to the same mRNA after the preceding one has moved out of the way gives rise to a polyribosome or polysome (see Figure 30). Translation continues until a stop codon is encountered, at which point the ribosome dissociates and releases both the mRNA and the completed protein. While translation is underway, up to 20 amino acids are added each second. Picture what this entails! Twenty times each second, a tRNA carrying the amino acid needed next is identified and shuttled into the ribosomal structure, a new covalent bond links the amino acid to the growing chain, the now empty tRNA exits the

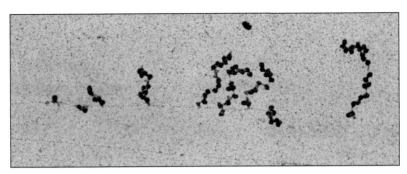

Figure 30: TEM of a polyribosome showing an mRNA with several translating ribosomes associated. The ribosomes are translating the mRNA from left-to-right in this image, as indicated by the relative lengths of the growing protein chains attached to each one. Courtesy of O. L. Miller, Jr., University of Virginia, and B.A. Hamkalo, University of California, Irvine.

ribosome, and the growing protein molecule is ratcheted one notch further into the cytoplasm. Accuracy? Only one error in 1,000 to 10,000 amino acids.

Reading Frame

The exact position where translation starts is critically important. Each mRNA presents a string of bases. For example:

...UCCAUACUAGAGUGCGUC...

Picture a ribosome viewing an mRNA through a window or reading frame that is three nucleotides wide. The three bases visible are interpreted as a codon, and then the next three are brought into view and likewise read as a triplet. The position of the reading frame demarcates each codon within the continuous string of bases, i.e., each word within the continuous string of letters. Depending on where the reading frame is placed, any base could be read as the first, second, or third base within a codon. That mRNA segment could be read as

...UCC AUA CUA GAG UGC GUC...
...Serine Isoleucine Leucine Glutamate Cysteine Valine...

or

...U CCA UAC UAG AGU GCG UC...

...Proline Tyrosine Stop Serine Alanine...

or

...UC CAU ACU AGA GUG CGU C...

Histidine Threonine Arginine Valine Arginine

There are three possible reading frames. Depending on the reading frame used, the first "A" in this sample segment is in the first, second, or third position within a codon. Thus the same base sequence encodes different amino acids when translation uses a different reading frame – the tactic that Yoda uses to encode more proteins in its short chromosome (see "Yoda" on page 41).

A mutation that substitutes one base for another will change, at most, one amino acid in the final protein (unless it introduces a stop codon, in which case the protein would be truncated at this point). A mutation that deletes or inserts nucleotides, however, often has a much greater effect. If the number of nucleotides added or removed is a multiple of three, then the change will be the addition or removal of one or more amino acids at this position in the protein. If it is not, the change shifts the reading frame and thereby affects every codon downstream of that location. The examples above show the results of first a +1 and then a +2 frameshift upstream of this segment. In both cases a completely different amino acid sequence resulted. Moreover, the +1 frameshift introduced a stop codon. Translation would stop at that point. When a frameshift eliminates a stop codon or other termination signal, a longer protein is synthesized.

The number of proteins of each type synthesized depends not only on the rate at which the corresponding genes are transcribed into mRNA, but also on how many copies of the protein are made using those mRNAs. Once synthesis of a protein has been initiated, the rate at which amino acids are added to the lengthening molecule is constant. To make more or fewer proteins, phages regulate the initiation step. Some mRNAs have a shorter half-life than others, which allows fewer proteins to be made before the mRNA is degraded. Folding of the mRNA (see "Nucleic Acid Structure" on page 56 and also "The Ingenious Minimalist" on page 92) can render ribosome binding sites less accessible, thus allowing ribosomes to bind and initiate translation less frequently. Although the folded structure depends first on the base sequence, it also varies depending on temperature or interaction with small molecules. This variation allows the quantity of proteins made to be fine-tuned in real-time in response to factors such as temperature and the presence of specific metabolites.[73]

Phages also know how to make two different proteins from one gene. One trick is to encode two ribosome binding sites in the mRNA. Depending on where the ribosomes attach, translation yields a mixture of a longer and a shorter protein that differ at their upstream end. By combining this with different accessibility of the two binding sites, the phage can tweak the relative quantities of the two proteins (see "Scoring a Hole-in-One" on page 172). Another tactic creates two different proteins by occasionally ignoring a stop codon (or other translation termination signal). Yet another way to do this is to encode a "slippery" sequence that causes the ribosome to occasionally lose traction and slip one or two nucleotides either forward or backward–a programmed frameshift. The net result of this maneuver is two proteins with different capabilities synthesized in markedly different relative numbers.

Precision in protein synthesis matters. Errors can be made during either transcription or translation. Due to the redundancy of the genetic code, many errors have no effect on the final protein. A transcription error that yields CCG instead of CCA is one example. In fact, CCG, CCA, CCT, and CCC all specify the same amino acid–proline. Likewise, the same error made during translation by a tRNA-mRNA mismatch would be "silent." Some errors that do result in a single amino acid substitution may have no detectable effect. The codon assign-

---

[73] metabolite: any of the many small molecules that participate in or are produced by cellular metabolism.

ments in the genetic code are such that often such errors replace one amino acid with a chemically similar one. For example, an error substituting GAG for GAC would substitute one acidic amino acid for another, specifically glutamic acid for aspartic acid. Scattered throughout a protein are specific amino acids that are essential for protein function, such as those at a reaction center where enzymatic catalysis takes place. Some others may be necessary for the protein to bind to another macromolecule, as in the case of a repressor that binds to a specific DNA sequence. More generally, an amino acid can be pivotal for folding of the protein into the three-dimensional structure required for it to function. Although substitutions at critical points can be deleterious, most amino acid substitutions are tolerated at many positions in these large macromolecules.

Nucleic Acid Structure

The base sequence comprises the primary structure[74] of nucleic acid molecules.

Based on this primary structure, both single-stranded and double stranded molecules can build three-dimensional structures that are stabilized by intra-molecular or inter-molecular base pairing – secondary structures.[75] Intermolecular base pairing gives rise to the familiar double helix formed by both dsDNA and dsRNA.

Single-stranded molecules build on their secondary structure to yield more diverse forms that fulfill specific functions. Consider the essential three-dimensional structure adopted by all tRNAs (see Figure 29).[76] When interacting with a ribosome, all tRNAs must align their anticodon to the mRNA codon while simultaneously placing the carried amino acid in position for linkage to the growing protein chain. Viewed in two dimensions, all tRNAs have similar cloverleaf structures with two ends and three "arms." Each arm is a stem-loop pro-

---

[74] primary structure: the base sequence in a nucleic acid or the amino acid sequence in a protein.

[75] secondary structure (nucleic acid): the three-dimensional structure of a DNA or RNA molecule that results from the formation of weak bonds (e.g., hydrogen bonds) between complementary bases within the same molecule or with an associated nucleic acid molecule. Examples: double helix, stem-loop.

[76] A short animation illustrating the folding of a tRNA molecule into its characteristic cloverleaf structure is available at http://bit.ly/2jFXayi.

duced when the ssRNA folds back on itself and forms a base-paired stem ending in a single-stranded loop. Two of these arms interact with the ribosome to position the tRNA for amino acid delivery. One arm presents the anticodon to the mRNA. The two ends of the molecule base pair to each other to form an open ended stem for attachment of a specific amino acid. All of these secondary structures in turn constrain the overall three-dimensional shape of the molecule by forcing the large-scale bends indicated in the diagram. This L-shaped arrangement locates the amino acid close to the nascent protein, while simultaneously positioning the anticodon in a distant stem-loop for matching to the codon in the mRNA. The molecule as a whole is structured to associate with the ribosome long enough to deliver its amino acid and then to quickly leave, allowing the next tRNA in.

Folding and base pairing of Minimalist's ssRNA chromosome produce similar stem-loops as well as unpaired "bubbles" (see "The Plight of RNA Chromosomes" on page 83). More complex three-dimensional architecture is possible here because this is a much longer molecule: 4,215 nt versus less than 100 for a tRNA. This chromosome structure embodies highly sophisticated mechanisms that not only allow different rates of translation for individual proteins, but also vary those rates as the infection proceeds (see "The Ingenious Minimalist" on page 92).

Protein Structure

Proteins are large macromolecules that are typically composed of several hundred amino acids, sometimes even thousands.[77] The final folded shape of any protein is critical for its function, and likewise its structure can provide us with clues as to its function. Many enzymes are globular proteins that bury the active site in the interior, often shielded from water. At the other extreme are the elongated structural proteins that form structures such as virion tail fibers. One tactic, used by Stubby, among others, to produce a fiber is to align three of these molecules side-by-side and twist them around each other, like the strands of a rope (see "Three Phage Tales" on page 133).

---

[77] Phages, more often than their prokaryote hosts, also make small proteins of less than one hundred amino acids.

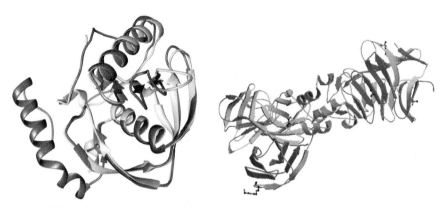

Figure 31: Protein structures. In ribbon diagrams, spiral regions indicate α-helical secondary structure, flat arrows denote β-strands, while the connecting threads are random coil loops. (Left) An overlay of two peptide deformylase enzymes showing that both have very similar structure and both bury their active site in the interior of the folded protein. Previously published in *Life in Our Phage World* by Rohwer, F, M Youle, H Maughan, N Hisakawa. 2014. Wholon. Used with permission. (Right) Stubby constructs its sturdy tail fibers like a rope, i.e., by aligning three protein molecules and then twisting the three "strands" around each other. Courtesy of RCSB Protein Data Bank (PDB 4a0t). Original publication: Garcia-Doval, C, MJ van Raaij. 2012. Structure of the receptor-binding carboxy-terminal domain of bacteriophage T7 tail fibers. Proc Natl Acad Sci USA 109:9390-9395.

Primary structure in proteins refers to the sequence of amino acids along the protein chain, while protein secondary structure[78] results from hydrogen bonds formed between the atoms engaged in the peptide bonds along the molecular backbone. The three most common secondary structures are the α-helix, β-sheet, and random coil, all of which can be represented in a ribbon diagram (see Figure 31). Interactions between the specific amino acid side chains contribute to the overall shape of the protein, i.e., its tertiary structure.[79] There is also a fourth level of structure, quaternary structure,[80] that results from the non-covalent association of two or more protein chains.

---

[78] secondary structure (protein): the three-dimensional structure of a protein molecule that results from hydrogen bonds between the atoms engaged in the peptide bonds along the backbone. The three most common secondary structures are the α-helix, β-sheet, and random coil.

[79] tertiary structure (protein): the overall shape of a correctly-folded, functional protein molecule that is superimposed on the regions of secondary structure that are, in turn, derived from interactions between the amino acid side chains.

[80] quaternary structure (protein): the arrangement of two or more protein molecules that associate through non-covalent bonds. Examples: the RNAP holoenzyme, Lander's long tail fibers.

Now turn the page. The phages are waiting for you.

Further Reading

Ackermann, H-W. 2007. 5500 Phages examined in the electron microscope. Arch Virol 152:227-243.

Avery, OT, CM MacLeod, M McCarty. 1944. Studies on the chemical nature of the substance inducing transformation of pneumococcal types induction of transformation by a desoxyribonucleic acid fraction isolated from pneumococcus type III. J Exp Med 79:137-158.

Brüssow, H. 2009. The not so universal tree of life or the place of viruses in the living world. Philos Trans R Soc B 364:2263-2274.

Forterre, P. 2013. The virocell concept and environmental microbiology. ISME J 7:233-236.

Hershey, AD, M Chase. 1952. Independent functions of viral protein and nucleic acid in growth of bacteriophage. J Gen Physiol 36:39-56.

Kausche GA, E Pfankuch, H Ruska. 1939. Die Sichtbarmachung von pflanzlichem Viren im Ubermikroskop. Naturwissenschaften 27, 292–299.

Kropinski, AM, D Prangishvili, R Lavigne. 2009. Position paper: The creation of a rational scheme for the nomenclature of viruses of Bacteria and Archaea. Environ Microbiol 11:2775-2777.

McLaughlin Jr, RN, FJ Poelwijk, A Raman, WS Gosal, R Ranganathan. 2012. The spatial architecture of protein function and adaptation. Nature 491:138-142.

Moreira, D, P López-García. 2009. Ten reasons to exclude viruses from the tree of life. Nat Rev Microbiol 7:306-311.

Pietilä, MK, TA Demina, NS Atanasova, HM Oksanen, DH Bamford. 2014. Archaeal viruses and bacteriophages: Comparisons and contrasts. Trends Microbiol 22:334-344.

Pina, M, A Bize, P Forterre, D Prangishvili. 2011. The archeoviruses. FEMS Microbiol Rev 35:1035-1054.

Raoult, D, P Forterre. 2008. Redefining viruses: Lessons from Mimivirus. Nat Rev Microbiol 6:315-319.

Rohwer, F, M Youle, H Maughan, N Hisakawa. 2014. *Life in Our Phage World*. Wholon.

Williams, TA, PG Foster, CJ Cox, TM Embley. 2013. An archaeal origin of eukaryotes supports only two primary domains of life. Nature 504:231-236.

Wolkowicz, R, M Schaechter. 2008. What makes a virus a virus? Nat Rev Microbiol 6:643.

Yokoo, H, T Oshima. 1979. Is bacteriophage φX174 DNA a message from an extraterrestrial intelligence? Icarus 38:148-153.

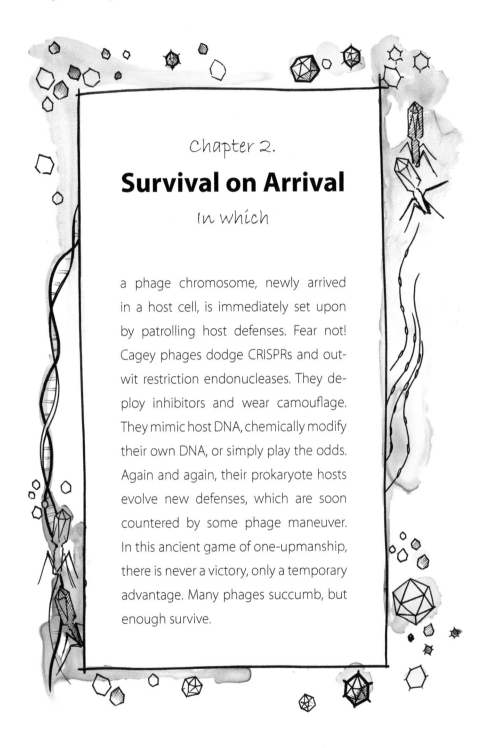

Chapter 2.

Survival on Arrival

In which

a phage chromosome, newly arrived in a host cell, is immediately set upon by patrolling host defenses. Fear not! Cagey phages dodge CRISPRs and out-wit restriction endonucleases. They de-ploy inhibitors and wear camouflage. They mimic host DNA, chemically modify their own DNA, or simply play the odds. Again and again, their prokaryote hosts evolve new defenses, which are soon countered by some phage maneuver. In this ancient game of one-upmanship, there is never a victory, only a temporary advantage. Many phages succumb, but enough survive.

Being a successful bacterium is not a secure occupation.
Ben Knowles

Growing up, I was taught that a man has to defend his family.
When the wolf is trying to get in, you gotta stand in the
doorway.
B. B. King

"Now, here, you see, it takes all the running you can do, to
keep in the same place. If you want to get somewhere else,
you must run at least twice as fast as that!"
The Red Queen in *Through the Looking Glass* by Lewis Carroll

The world is but a perpetual see-saw.
Michel de Montaigne

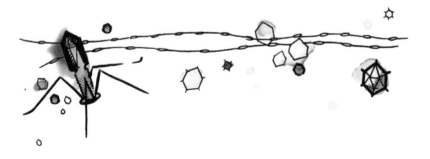

E very infection starts with the successful delivery of a phage chromosome into the right type of prokaryotic cell, i.e., a member of the particular strain or species that this phage knows how to exploit. However, every successful delivery does not result in a successful infection. Bacteria[1] are vigilant, with anti-phage defenses always deployed and poised to destroy any invader. It would do no good to argue with the bacterium, to point out that phage infection was essential for the evolution of its species or that the phages continue to drive bacterial diversity and enable Bacteria to flourish in balance with their environment. At the moment, this particular phage invasion is a matter of life and death for that particular bacterium. The cell responds to secure personal survival, if it can, by destroying the would-be invader before it can launch an infection. The phages, in turn, have no choice but to continually scramble to counter their host's latest defenses. The advantage sea-saws back and forth, between phage and host, with no checkmate possible.

Death by Nuclease

Diverse nucleases[2] patrol the cytoplasm of every bacterial cell. Some of these enzymes cut RNA molecules while others attack DNA; some start hacking at a free end of the molecule while others cut internally; some recognize and cleave at specific sequences while others are indiscriminate. You might wonder, if the invading phage chromosome is dsDNA, like the chromosomes of all cells, how does a cell selectively target the phage DNA without attacking its own chromosome? This ability to discriminate between self and non-self is the essence of immunity—a skill mastered not only by animals, but also by plants, fungi, and even Bacteria. This self-assessment can be simple for a bacterium. Typically its own chromosome is a circle of dsDNA, whereas a phage dsDNA chromosome arrives as a linear molecule. Having ends renders a phage chromosome vulnerable to attack by cytoplasmic exo-

---

[1] Although phages *sensu lato* include viruses that infect Archaea as well as Bacteria, I will usually refer to hosts in general as Bacteria. The hosts in the vast majority of the specific stories that follow are, in fact, Bacteria, because we have barely begun to explore the tactics of the archaeal viruses.

[2] nuclease: an enzyme that cleaves a molecule of DNA or RNA. Endonucleases cut internally; exonucleases clip off nucleotides from the ends of a linear molecule.

nucleases that would quickly clip off nucleotides, one by one, from those ends. However, the exonucleases have to strike quickly, as most phage chromosomes circularize as soon as the entire chromosome is inside the cell. The exonucleases don't get a second chance because the phage chromosomes remain circular throughout the infection, resuming linearity only as they are packaged inside a capsid. Sometimes the exonucleases don't even get a first chance. A few phages, such as Lander, deliver a protein[3] along with their chromosome that binds to the chromosome ends and shields them from attack.

Once the phage chromosome becomes circular, the invaded cell needs another criterion to distinguish the invader from its own DNA. Many Bacteria make a DNA methyltransferase[4] and use it to routinely tag their own chromosomes. This enzyme attaches a methyl group[5] to selected cytosines or adenines after they are incorporated into the DNA. Since the added methyl groups tuck unobtrusively into one of the grooves in the double helix, they do not block base pairing, transcription, or DNA replication. Not all cytosines and adenines are modified this way, only those that reside within the specific base sequence recognized by that bacterium's specific DNA methyltransferase. All such qualifying sites are methylated on both DNA strands. These recognition sites are typically palindromes[6] composed of 4–8 nucleotides. Already thousands of these methyl-tagging enzymes with more than 2,000 different recognition sites have been identified in Bacteria.

Palindromic Recognition Sites

Palindromes in our written languages have been a source of entertainment. Here a palindrome is a sequence of letters that are the same when read in opposite directions, as in: amanaplanacanalpanama.

By analogy, a palindrome in a double-stranded nucleic acid molecule reads the same on one strand as when read in the op-

---

[3] Lander: enterobacteria phage T4, a myovirus that infects *E. coli*. For Lander, this protein is gene product 2 (gp2).

[4] DNA methyltransferase: an enzyme that transfers methyl groups from a donor molecule to specific bases in DNA.

[5] methyl group: –CH3, i.e., 1 carbon atom with 3 attached hydrogen atoms.

[6] palindrome: in a double-stranded section of a DNA or RNA molecule, the presence of the identical sequence in both strands when they are read in opposite directions.

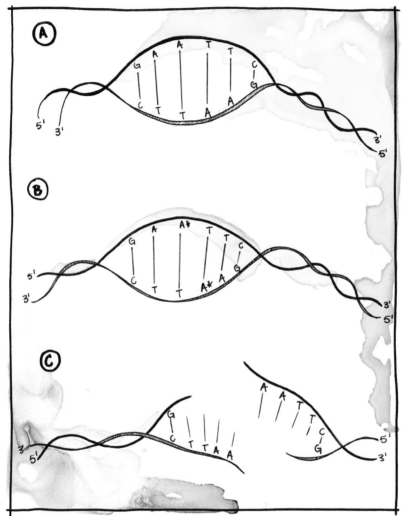

Figure 32: Palindrome Recognition Sites

posite direction on the complementary strand. For example, this six base pair segment of DNA (see Figure 32) is a palindrome that reads GAATTC on the upper strand from the 5′ end[7] toward the 3′ end and also on the complementary lower strand

[7] The 5′ and 3′ designations refer to the carbon atoms in the five-carbon sugar molecule in each nucleotide. These are the sites where each sugar is linked to its neighboring phosphates in the nucleic acid backbone. During synthesis, nucleotides are added at an available 3′ carbon. The first nucleotide in the chain has an unused 5′ site, and the last has an unused 3′ site. Every DNA or RNA molecule has a directionality, and two complementary strands have opposite orientations.

when read from 5′ to 3′. This particular palindrome is recognized by one of *E. coli*'s DNA methyltransferases that adds a methyl group to the 3′ adenine indicated by the asterisk (B). If that adenine is unmethylated, the host's cognate[8] restriction endonuclease (see "Restricted Zone" on page 67) will cut each strand between the G and A residues, which leaves complementary single-stranded overhangs on both strands (C).

Restricted Zone

What threatens the phages is the enzyme that shadows each of these site-specific DNA methyltransferases. This second enzyme identifies the same recognition site in DNA. Moreover, it can tell whether or not that palindrome bears the specific methyl tag that certifies the DNA to be the cell's own. If it encounters dsDNA that is not tagged correctly on at least one strand, it pronounces the DNA foreign and immediately cleaves both strands at or near the telltale sequence. If that DNA was an invading phage chromosome, that phage is dead. These murderous enzymes are known as restriction endonucleases[9] (REs) – *restriction* because they restrict phage replication and *endonuclease* because they make internal cuts in a nucleic acid. A DNA methyltransferase together with its cognate RE comprise a restriction-modification system – one bacterial way of converting a would-be invader into a nutritious lunch.[10]

DNA Methyltransferases: More Than an Antiphage Defense

Bacterial DNA methyltransferases do more than protect the cells from foreign DNA. Some, such as DAM[11] in *E. coli*, have no cognate RE and serve a different function. DAM cor-

---

[8] cognate: (adj.) a term borrowed from linguistics and used to indicate some correspondence between two molecules. In this case, a DNA methyltransferase and its cognate restriction endonuclease recognize the same methylation pattern in the same palindrome.

[9] restriction endonuclease: (RE) an endonuclease that recognizes specific sites in DNA, typically 4–8 nucleotides long, and cleaves the DNA if the site lacks the correct pattern of base methylation.

[10] In some cases the methyltransferase and RE are parts of the same multifunctional protein, but in other cases they are independent proteins, as described here.

[11] DAM: **D**NA **a**denine **m**ethyltransferase.

rects mistakes made during chromosome replication. The DNA polymerase holoenzyme zips along at a remarkable speed, adding 500 nucleotides or more to a growing DNA strand each second. An incorrect base is inserted only once in every 200,000 to 2,000,000 (2×10^5 to 2×10^6) base pairs. That error rate sounds quite good by our industrial standards, but it isn't good enough for a bacterium. It would mean that for every bacterial chromosome containing approximately 4,000,000 base pairs, 2–20 errors would be introduced in each new chromosome. The DNA polymerase itself proofreads and corrects many of its own mistakes, reducing the error rate to only one per every 8,000,000 to 400,000,000 (8×10^6 to 4×10^8) base pairs, or only one error in every 2–100 chromosomes. (The wide range reflects different rates for different types of base substitutions.)

Better accuracy is provided by adding yet another level of error-checking, this one involving DAM. Recall that DNA replicates semi-conservatively to yield a new daughter chromosome composed of one parental strand and one newly synthesized strand. Insertion of an incorrect base in the new strand produces a mismatch between it and the parental strand. For example, instead of the correct pairing of a G in one strand with a C in the opposite strand, a G might be opposite a T. In order to correct this mistake, the bacterium needs to not only recognize the discrepancy, but to also know which strand is correct—in other words, to know which strand is the parental strand. DAM tags the bacterial DNA by adding a methyl group to the adenine in every GATC sequence. Significantly, it does not do this during replication, but instead shortly afterwards. Thus, for a brief period right after synthesis, the GATC sequences in the parental strand are methylated, but those in the new strand are not. This is the clue that tells the mismatch repair system which base to replace, and which one is correct.

This third level of quality control reduces the error rate by another factor of 20–400 to yield at most one error in every 40 chromosomes – an acceptable rate. Might Bacteria, given another billion years, reduce this rate even farther? I doubt it. I would argue that they already, long ago, achieved an optimal balance that provides adequate variation through mutation for natural selection to act on, while maintaining an adequate level of replication fidelity. Striving for greater accuracy would cost additional cellular equipment and thus would consume more energy, but without a payback in increased fitness.

Phage chromosomes are not the only foreign DNA or RNA seeking entry into a bacterium. There are also other mobile genetic elements[12] – plasmids, transposons, and mobile genetic islands – all of which are genetic information on the move without the protection of a protein capsid. Moreover, some Bacteria actively invite the entry of foreign DNA. These competent[13] Bacteria take up DNA fragments from their environment, but typically only a small minority of the population do, and only when they are starving or otherwise stressed. Foreign DNA is a good thing for a cell provided it doesn't try to takeover. Most of it is quickly degraded by cell nucleases and its building blocks recycled. This can provide a significant source of nutrients in some environments. Far less often a segment of the incoming DNA is incorporated into the host chromosome by recombination. Likewise, the DNA fragments produced when an incoming phage chromosome falls prey to the REs are also available for recombination into the host chromosome. The acquisition of foreign DNA by these routes is infrequent, but proof of its occurrence is apparent in every bacterial chromosome. This role of phage in bacterial evolution will be explored in PIC.

Evading Restriction

Restriction is widespread, plaguing the phages wherever they go. This form of innate immunity is deployed by 90% of the Bacteria studied thus far. Despite its great effectiveness, it remains inherently a leaky defense. Some phages do slip past. DNA methyltransferases and REs

---

[12] mobile genetic element: a molecule of DNA that typically encodes one or more proteins and that can move between prokaryotic cells or between locations on the chromosome(s) within a cell. These include plasmids, transposons, etc.

[13] competence: in Bacteria, the ability to take up DNA from the environment. Most often this uptake is activated during starvation and the incoming DNA is recycled. However, some Bacteria selectively take up DNA from close kin and occasionally recombine the fragments into their own chromosome.

both monitor the host cytoplasm, continually perusing any DNA encountered and evaluating the methylation status of their specific recognition sites. Usually the REs outnumber the DNA methyltransferases, but there is still the possibility that all unmethylated sites in an invading phage chromosome will be recognized by a DNA methyltransferase first. When that happens, the enzyme methylates each site, thereby shielding it from restriction. If the RE encounters a site first, it strikes. It's a probability game as to whether the phage lives or dies. The odds are against phage survival. Protection requires methylation of every target sequence, whereas restriction requires only a single strike. Nevertheless, often one phage in a hundred thousand or one in a million escapes restriction by becoming fully methylated, and that escapee can win big.

If the infection attempt is successful, the original, but now methylated, phage chromosome will be replicated to yield two new chromosomes, each with one methylated and one unmethylated strand – the same as the cell's own DNA immediately following its replication. As long as one strand is methylated, the DNA is safe from RE attack. The new strand soon acquires its methylation and replication continues unmolested by the REs. When the infection ends, all of the progeny virions emerge from the host fully methylated. Nearby Bacteria related to that host are apt to carry the same restriction-modification system. If these phage progeny infect one of them, the patrolling REs would find the incoming DNA to be appropriately methylated and would erroneously certify the invader to be "self." This new infection would proceed, and once again numerous progeny virions would go forth into the world camouflaged with this host's pattern of methylation. These generations of methylated phages escape restriction for a while, but only until they attempt to infect a cell with an RE that has a different target specificity. Then once again their chromosome is lunch.

Rather than rely on this improbable escape route, some phages encode "orphan" DNA methyltransferases originally acquired from their host.[14] In typical phage fashion, these phages not only pinched the genes from their hosts, but they then modified them to better suit

---

[14] orphan DNA methyltransferase: a DNA methyltransferase that is not accompanied by a cognate restriction endonuclease (RE).

their own purposes, in this case adding versatility. Some of these orphans methylate two, or even three, different target sequences, rendering them invisible to host REs that monitor any of those sequences for appropriate methylation.

A phage can also improve its chances of survival by reducing the number of targets in its chromosome. The more restriction sites a phage carries, the more likely it is that the chromosome will be cleaved before all sites are methylated. Is this dodge practical? Yes, indeed. Knocking out a target requires only a single point mutation[15] anywhere in a targeted palindrome. Given the number of phages replicating every minute, such rare events occur often in the phage population (see PIC). Frequently these simple replacements have no deleterious effect on the phage. The genetic code has considerable redundancy (see "DNA Genetic Code" on page 47). Often a single base can be replaced in a protein-coding gene without altering the amino acid sequence of the protein. Even when the base substitution results in an amino acid substitution, this may have little, if any, effect on protein function. When a phage chromosome contains several vulnerable restriction sites, eradicating even one improves the odds for escape from RE attack. This advantage will be inherited by the next phage generation. They, in turn, have the opportunity to jettison another target site and further improve their chances for survival.

Bacteria also play the probability game. A bacterium armed with two or more REs that target different sequences is more likely to zap an invader. Suppose only one phage in a thousand slips past one of those REs (one in 10^3), while only one in ten thousand (one in 10^4) escapes the second RE. Having both REs on patrol would reduce phage survival to only one in 10^7– much better odds for the bacterium. The Bacteria definitely take advantage of this tactic. E. coli alone has acquired a roster of more than 600 REs monitoring many different target sequences. Different strains may carry several different ones, and an incoming phage never knows which ones will be there, waiting. No phage can eliminate every one of those recognition sites. And even if they could, the bacterium would soon acquire or evolve an RE with a different target and be one step ahead of the phage again.

---

[15] point mutation: the replacement of one base in a chromosome with another.

Instead of scrambling to evade each particular RE, a thinking phage would realize the advantages of dodging all bacterial REs regardless of their specific target sequence. This, in fact, is Stubby's[16] strategy. Successful implementation is tightly coupled to several other steps in Stubby's infection cycle. For instance, during virion assembly each capsid is filled with an identical linear chromosome positioned so that the same chromosome end will exit the capsid first. When a suitable *E. coli* host has been contacted, Stubby delivers its chromosome slowly – 12 minutes from start to finish. For the first few minutes, the portion already inside the cell is protected by some not yet identified mechanism. However that is done, it does not prevent transcription of the genes in that region, and their translation follows immediately. Soon Stubby has synthesized and deployed its first protein – its endonuclease inhibitor. This protein can mimic a generic segment of DNA, regardless of its base sequence. Two of these proteins couple together to form a homodimer that imitates the shape and electrical charge distribution of a DNA helix when it is bent in the clutches of an attacking RE. This decoy does fool the REs, but that alone would not be sufficient to inactivate them all. These dimers also bind so avidly to the enzyme that they can outcompete the DNA for occupancy of the RE's active site. To occupy, and thereby neutralize, all of the patrolling copies of one common RE, Stubby needs to make only 10–40 dimers – a task that it quickly accomplishes soon after the first portion of its chromosome has entered the cell. Although very effective, this inhibitor has its limitations. There are several different classes of REs, and some of them aren't fooled.

A phage such as Yoda[17] that uses ssDNA for its chromosome does enjoy some immunity since REs cleave only dsDNA, but this protection is short-lived and also incomplete. Yoda's circular chromosome is indeed all single-stranded as it enters the cell, but that soon changes. There are many complementary sequences within the chromosome that could base pair with another region of the same DNA strand (i.e., intramolecular base pairs). As much as a third of the chromosome might become vulnerable this way once the entire chromosome is inside the cell. Later in the infection this vulnerability decreases. Once the synthesis of

---

[16] Stubby: enterobacteria phage T7, a podovirus that infects *E. coli*.
[17] Yoda: coliphage ϕX174, a microvirus that infects *E. coli*.

abundant capsid proteins is underway, these proteins shield the new ssDNA chromosomes by condensing into a capsid around them (see "The Yoda Way" on page 142). Of more serious concern, soon after arrival in the cell, Yoda's entire chromosome seemingly becomes susceptible to RE attack because the first step in chromosome replication is the formation of a dsDNA replication intermediate. Yoda's strategy for protecting this double-stranded form is unknown, but effective. This phage produces more than a hundred progeny from each infection and has earned a place in the competitive phage world.

The strong incentive to escape restriction has also driven the evolution of some highly sophisticated solutions that involved major changes to phage DNA which, in turn, required entirely new enzymatic functions. Some phages replaced one of the usual four bases (G, C, A, and T) with some variant, such as swapping 5-hydroxymethyluracil for thymine (T) or replacing every cytosine (C) with a modified cytosine. Most REs are thwarted when they encounter such alterations. Lander's ancestors exploited this tactic. They started with the traditional cytosine and then modified it, step-by-step, earning a period of respite each time, but each respite lasted only until the host evolved a countermeasure. This evolutionary sparring continues today – a perpetual arms race.

The Saga of Sweet Lander (*A soundly-based historical reconstruction*)

Once upon a time, long ago, *E. coli* cruised confidently through phage-infested waters. Among its strains could be found three different classes of REs that collectively monitored methylation of hundreds of different target sequences. Nevertheless, Lander's ancestors could eke out at least a meager existence because restriction is never absolute. The scales tipped to favor the phages when those phages evolved a way to foil all of the many REs that check for an unmethylated cytosine. Their method? Methylate every cytosine in their chromosome. This preemptive tactic was effective until *E. coli* countered with a new class of REs that specifically destroyed DNA that contained methylcytosine in their target sites.

Now the phages were faced with some REs that targeted cytosine and with others that struck methylated cytosine. However, a phage could evade them all–for a while–if instead it used exclusively 5-hydroxy-methylcytosine (HMC) when synthesizing its DNA. This may sound like a simple alteration, but it wasn't. Granted, HMC base pairs with G just like C or methylcytosine, but there is also a pool of cytosine available in the host cytoplasm. The phage had to insure none of it was used instead of HMC when replicating its DNA. This required diligent employment of new enzymes to convert all the cytosine to HMC. In addition, since most DNA polymerases won't accept HMC as a substitute for cytosine, the phage also evolved a new DNA poly-merase that would. The net result was phage DNA that contained HMC and no cytosine. Now the only DNA in the cell that contained cytosine was host DNA – a circumstance that the phages proceeded to exploit mercilessly. They set their own endonucleases to work to se-lectively degrade the host's cytosine-containing chromosome and then recycled those nucleotides into phage DNA. They also modified the host's RNA polymerase to preferentially transcribe HMC-containing phage DNA, thereby redirecting it away from host genes and rapidly shutting down synthesis of host proteins.

The impressive immunity gained by HMC was, as you would expect, only temporary. Over time, *E. coli* (strain K-12) evolved an enhanced restriction system that specifically cleaved HMC-containing DNA. Un-der this new attack, the phages responded with sweetness. After each HMC was incorporated into the DNA, the phages adorned it with a sugar molecule. Lander added glucose, while some of its kin used oth-er sugars. These phages with sugar-coated DNA were immune to even the REs that specialized in HMC-containing DNA.

The contest did not stop there. Some pathogenic strains of *E. coli* (e.g., CT596) found a way to destroy the sweetened DNA. Building on the existing REs, these strains developed a whole new class of these en-zymes, each one adapted to cope with DNA carrying a particular sug-ar attached to the HMCs. Furthermore, the useful genes for these new REs were transferred from one host strain to another by mobile genetic elements. Soon this capability spread throughout the *E. coli* popula-tion – bad news for Lander.

Some of Lander's relatives are still stumped by this new defense, but not Lander. It packages a protective shield inside each new virion along with the chromosome – approximately 360 proteins that inhibit *E. coli*'s sugar-loving REs.[18] These inhibitors include several variants, each one targeting a slightly different host RE. They are, of necessity, small, compact proteins ($30 \times 30 \times 15$ Å), so compact that during DNA delivery they can pass intact through the narrow capsid exit portal.

Already this latest phage defense is starting to erode. Two pathogenic strains of *E. coli* now sport modified REs that are immune to these inhibitors. It is only a matter of time before this salutary bacterial innovation spreads to other strains. Rest assured, though, that it is also only a matter of time before Lander comes up with an effective countermove. Even now the first mutants able to escape these latest REs have shown up. We don't yet know how they manage their evasion. The future? This chess game will continue. Always a check, never a checkmate.

CRISPR Surveillance

REs are not the only endonucleases poised and waiting to destroy an incoming phage chromosome, nor are they the most sophisticated antiphage weapon in the prokaryotic arsenal. Restriction is a type of innate immunity that defends prokaryotes against any invading DNA, so long as it is recognized as non-self. Innate immunity encompasses a variety of primary defense mechanisms that are found in virtually all organisms, while the more "advanced" adaptive immunity was long believed to be the province of only "higher" organisms such as vertebrates. However, both Bacteria and Archaea have developed a highly phage-specific defense that is a form of adaptive immunity. What is the key difference between innate and adaptive? Adaptive immunity exploits the ability to remember previous infections and to then mount a more rapid and effective response if that same entity trespasses again. About 40% of the Bacteria and some of the Archaea whose genomes have been sequenced have adaptive immunity in the form of a CRISPR system.[19] The bacterial version will be featured here since we know much less about the archaeal mechanism.

---

[18] internal protein 1 (IP1).
[19] CRISPR: (pronounced "crisper") Clustered Regularly Interspaced Short Palindromic Repeat.

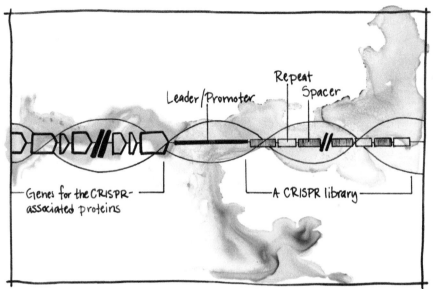

Figure 33: A CRISPR Locus

These "simple" bacterial cells do indeed remember specific phages (and other mobile genetic elements) that they previously confronted. This memory resides in their CRISPR library, a historical catalog of fingerprints of phages met by this particular cell's predecessors, thus of phages that are likely to also be in the current local population (see Figure 33). To add a phage to that catalog, the bacterium acquires a short segment from the phage chromosome, often in the range of 25 to 40 or 50 bases. This sample, called a spacer, can be taken from either strand of an invading dsDNA phage chromosome, from any gene, from any location provided it is adjacent to a short (2–8 nt) recognition motif called the PAM (**p**rotospacer **a**djacent **m**otif). The phage-specific spacer is added to the library without its adjacent PAM, but separated from its neighboring spacers by a short palindromic repeat sequence (the source of the "SPR" in the CRISPR acronym). New spacers are always added at one end of the library. If you read through the library starting at that end, you travel back in time from recent events to ancient history. These cellular archives can be extensive. Some libraries contain several hundred spacers, and some Bacteria have more than one library. To keep the size of each library in check, some spacers are deleted from time to time; to keep the collection current, older spacers are preferentially removed.

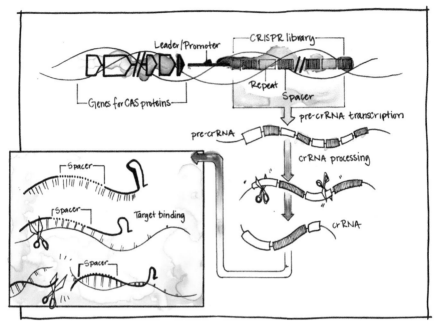

Figure 34: CRISPR Patrol

A CRISPR defense never sleeps. The cell continuously transcribes its entire CRISPR library to produce long RNA molecules containing all the spacers (a pre-crRNA; see Figure 34). The adjacent leader region contains the promoter[20] that initiates transcription of the repeats and spacers downstream. This RNA is then processed into short chunks, one spacer per chunk. These chunks, known as CRISPR RNAs, team up with an endonuclease to monitor all dsDNA in the cell – host and invader alike – looking for the matching spacer sequence with an adjacent PAM. Since the spacers in the library lack PAMs, they are ignored. When a matching spacer is detected next to a PAM, as in a phage chromosome, the endonuclease makes a lethal, double-stranded cut. One cut, and one phage is stopped in its tracks.

The diverse Bacteria and Archaea have evolved an assortment of CRISPR systems all of which follow the same general pattern. While the details can vary significantly from one bacterial strain to another, each contains a spacer library along with the **CRISPR-as**sociated proteins

---

[20] promoter: a region in the DNA upstream of the transcription start site that is recognized by the RNAP holoenzyme and promotes initiation of transcription of the DNA downstream.

(CAS proteins) that do the work of spacer acquisition, CRISPR RNA processing, and phage destruction. Likewise, in every CRISPR locus the genes for the CAS proteins are located next to the library on the bacterial chromosome. Thus, this cluster of genes contains everything needed for a CRISPR defense. Sometimes an entire defense system is transferred as a gene cluster from one bacterium to another, a bacterial form of plug and play. Not only is the recipient handed a gun, but the gun is already loaded and precisely aimed.

How does a bacterium nab a segment of DNA from an infecting phage and live to pass it on to their progeny? It takes several minutes to insert the DNA segment into the library, transcribe the library, and process the new spacer into an active CRISPR RNA. By then, most likely the phage will already have taken control of the cell. But Bacteria have a way, a tactic that will sound familiar to you because we use it ourselves: vaccination. Our vaccines present a specific antigen or an attenuated strain of a pathogen to our immune system. This prompts an immune response and also stores a pathogen-specific record in cellular memory. If that same pathogen later attempts an invasion, our adaptive immune system can respond swiftly.

The "vaccines" used by Bacteria are phage chromosomes that were cleaved after entry or were irreparably damaged prior to delivery. Picture a bacterium that carries both an RE and a CRISPR defense. Suppose that a phage delivers its dsDNA chromosome into the cell only to have its chromosome cleaved by the patrolling RE. Worse yet for the phage lineage, the CRISPR system acquires a spacer from one of the chromosome fragments. When a similar phage later attempts a takeover, it is met by both the REs and a targeted CRISPR defense. Another opportunity for bacterial vaccination arises when an invading phage chromosome is dead on arrival. Although close to 100% of virions may be infectious when first released into the environment, not every virion remains infectious. While en route to a new host, every virion is subjected to numerous environmental hazards (see *"Hic Sunt Dracones"* on page 188). In marine and freshwater environments, or on exposed surfaces such as plant leaves, the ultraviolet (UV) component of sunlight can damage phage DNA inside the capsid. The rate of phage inactivation is strongly dependent on the intensity of the incident UV light, but numerous other environmental factors also come into play. Sometimes this damage is repaired after arrival courtesy of the host's DNA repair mechanisms, but other times not. Without repair, such chromosomes pose no threat to the cell but are another potential source of new spacers.

Poking Holes in the CRISPR Net

A CRISPR system is a superb defense, but it's not invincible. The phages have – of course! – found numerous ways to penetrate or exploit it. Some ways are highly idiosyncratic. For example, Lander exploits its modified DNA. HMC, either with or without a glucose attached, partially shields DNA from the CRISPR endonucleases. Another stratagem takes advantage of the precision of many CRISPR systems and the rapid rate of phage evolution. In order to identify a target, many CRISPRs require an exact match between the archived spacer and the invader, and also a perfect PAM sequence adjacent. To escape, a phage needs only a single base substitution in either sequence. Point mutations sound to us like rare occurrences, being in the range of 3–4 per thousand phage genomes, but they are frequent given that a hundred or more genomes can be produced by each infection and that, in the oceans alone, 10^{23} infections occur every second. When such a chance mutation happens to be in the right location to neutralize a host's CRISPR defense, the lucky phage and its progeny will thrive.

Some Bacteria raise the bar higher. Suppose phage P is abundant in their environment. A bacterium that already has one spacer targeting phage P is apt to acquire additional spacers against the same phage. Each time its CRISPR endonuclease is guided by that spacer to cleave an invading phage P chromosome, the bacterium has the opportunity to acquire another, different spacer from the fragments. As a result, an invading phage with an escape mutation that evades one spacer will still be halted by the other. Phage survival now requires mutations in both of the target sequences. If one phage in a million escapes from a single spacer, targeting by two spacers reduces the number of escapees to one in 10^{12}. Doubly-protected, this bacterium and its descendants will come to comprise a larger proportion of the local population, tipping the current advantage to the bacterium.

Not all CRISPR systems require an exact match. Some instead tolerate one or more base mismatches between the spacer and the phage. Here, in order to escape detection, a phage must mutate at least two nucleotides in that target sequence – a less frequent occurrence. Such tolerant CRISPRs provide another bonus to the bacterium. Closely related phages have many genes in common that they inherited from a common ancestor (homologous genes[21]). Often these genes contain regions with nearly identical sequences. Thus, given a mismatch-tolerant CRISPR defense, a spacer acquired from one phage may also defend against related phages. This imprecision effectively extends a CRISPR memory to recognize targets not previously seen.

Phages sometimes slip through another hole in the CRISPR net. They frequently modify their genome sequence by another method besides mutation. During an infection, they make 50 or even hundreds of copies of their chromosome. These copies have many opportunities to interact and exchange homologous sections. This homologous recombination[22] can be so frequent in a natural phage population that their chromosomes become mosaics in which adjacent segments come to have different evolutionary histories (see PIC). If such DNA swapping interrupts a region targeted by a CRISPR spacer, it is apt to create a mismatch that fools the spacer. The net result is one progeny phage with immunity to that spacer, and that phage can live on to successfully infect that bacterial lineage.

Escape is more difficult in natural populations. Picture a bacterium living in a lake or in forest soil. During good times it grows, divides to become two cells that grow and divide to yield four, and so on, to potentially form a localized cluster of 32 genetically identical cells, i.e., a clone.[23] Many such bacterial clones exist side-by-side in every population. Each one is likely to carry a different assortment of spacers, a testament to their different infection histories. A phage that would be

---

[21] homologous genes: genes present in two different types of phage (or in two different species of Bacteria or Archaea) that were inherited by both from a common ancestor.

[22] homologous recombination: the reciprocal exchange between two, usually homologous, DNA molecules. This serves as a mechanism for DNA repair, as well as for the incorporation of genes from a related DNA molecule.

[23] clone: a group of cells or organisms, such as Bacteria or Archaea, that are descended from a single ancestor and that are genetically identical.

destroyed when attempting to infect one might meet no CRISPR opposition if, by chance, it instead infects a closely related cell right beside it. Furthermore, even when many clones carry a spacer targeting the same phage, each clone may have a different spacer. Although a single point mutation may enable a phage to replicate in one clone, its progeny would likely be targeted by different spacers in other clones and be destroyed. Chance and probability affect the fate of individuals on both sides. Successful individuals survive and give rise to lineages with a temporary – but only temporary – survival advantage.

The Best Defense Is a Good Offense

Rather than play any of these sequence-dependent games, why not take the offensive and simply inactivate a critical component of the CRISPR machinery and be done with it? Some phages do exactly that. Because there are many different ways to disable a CRISPR defense, it is not surprising that at least nine distinct families of anti-CRISPR proteins are made by various phages. Some anti-CRISPRs disrupt one step, some another. Some, for example, knock out the CRISPR patrol by blocking formation of the CRISPR RNAs. Others allow the patrols to identify their phage target, but then render them harmless by impeding recruitment of the endonuclease to the scene.

Phages also go on the offensive by employing their own CRISPRs. One phage, when subject to lethal attack during infection of a host cell, uses its CRISPR system to target a critical gene in the host's defense system. Upon arrival, the phage immediately mobilizes its CRISPR system to make a quick endonuclease strike against that gene. Moreover, because this phage CRISPR system is also able to update its spacer library by spacer acquisition, the phage can keep pace with mutations in the target sequence. In summary, an effective, well-deserved payback.

Altruistic Suicide

Even when a phage chromosome has dodged all host defenses and its success seems assured, disaster can still strike. The cell can up and die – an apparent altruistic suicide. (Animals, including humans, employ the same tactic as part of their immune response to infection, here sacrificing a few cells for the sake of the many.) A phage can't produce progeny inside a dead cell. Granted, this sacrifice is an extreme defense, a medicine that cures an infection but leaves the patient dead. That many diverse Bacteria use this tactic suggests that it provides a survival advantage to the lineage, if not to the individual. How might this work?

Once again, picture a small bacterial clone, perhaps 16 cells, growing in a favorable environment. Along comes a phage that successfully infects one of them, replicates, and ultimately sends 25, 50, or more infectious virions out into the neighborhood to search for their own hosts. Suitable hosts, in the form of other members of that clone, are conveniently at hand, and the slaughter escalates. If instead that first infected cell had died before any phage progeny had been produced, the death of that one bacterium could have saved its nearby clone mates from fatal infection. In that case, the death of that cell would appear to a human observer to be an altruistic deed for the sake of the cell's siblings. There are now, in this simplistic example, 15 Bacteria alive and reproducing, all passing on to their offspring the same genes as were carried by the now dead cell, including those responsible for the suicide mechanism itself. Close relatives without the suicide defense would suffer more phage casualties. Given the continuing presence of phage, the suicide-capable lineage would come to dominate the local population. But is this act truly altruistic suicide?

Not necessarily. It could be a byproduct of a seemingly innocuous mutation in the bacterial genome. Consider the case of Yoda's host, *E. coli*. DNA replication by *E. coli* requires a helicase. This enzyme assists DNA polymerase by unwinding short regions of the DNA to provide the needed single-stranded templates. Different variants of this helicase, the products of past mutations, are used by different strains of *E. coli*, but they all function smoothly with the same DNA polymerase. For countless generations, Yoda (ϕX174) has relied on hijacking the host's helicase to assist with the packaging of its own chromosome within its capsid. Some helicase variants that work just fine for *E. coli* derail Yoda's infection. Yoda still replicates its chromosome, still synthesizes all the proteins it needs, and still lyses the cell, but no infectious progeny virions are released. What has gone wrong? This particular helicase variant doesn't function with Yoda's chromosome packaging machinery. As a result, no infectious virions are produced but the cell dies – giving the appearance of an apparent altruistic suicide. It is not necessary to invoke altruism to explain this observation. Natural selection is sufficient. As long as these phages are preying on *E. coli*, the gene for the phage-resistant helicase increases bacterial survival and will be favored by natural selection. Similar scenarios have evolved many times in diverse Bacteria, each time involving alteration of a different bacterial protein that inadvertently interferes with phage replication in its own particular way. As a result, any phage lineage repeatedly faces such hazards, each an example of bacterial one-upmanship that must be countered by a phage escape mutation.

The Plight of RNA Chromosomes

Phage chromosomes composed of dsRNA slip past all of the endonucleases described so far, both the REs and the CRISPR-associated ones, but there is another one poised to attack them. Lengthy molecules of dsRNA are not a normal constituent of bacterial cells. Any dsRNA chromosome detected would be categorically condemned as foreign and cut by a bacterial endonuclease, RNase III – a specific target sequence not required. Nevertheless, there are phages, such as Shy[24], that manage quite well, their dsRNA chromosomes notwithstanding. Shy's dodge is to not deliver its dsRNA naked into the cell (see "Shy's Delivery" on page 232), nor expose it to the RNase III at any point during its replication cycle. Shy arrives instead as an efficient, capsid-enclosed replication machine made up of three dsRNA chromosomes plus an RNA-dependent RNA polymerase (RdRP)[25] attached to the inside face of each capsid vertex. (For the story of how Shy packages one copy of each of three different chromosomes inside every virion, see "Shy's Packaging Feat" on page 145.) The RdRPs transcribe the chromosomes inside the capsid, and the ssRNAs produced exit to the cytoplasm. There some serve as mRNAs for synthesis of phage proteins while others are routed to production of new chromosomes for the progeny phage. Chromosome production poses two challenges. The new chromosomes must not be exposed to the RNase III endonucleases, and their synthesis from the ssRNAs requires the phage RdRP

---

[24] Shy: phage φ6, a cystovirus that infects Pseudomonas species, primarily plant pathogens.

[25] RNA-dependent RNA polymerase (RdRP): an enzyme that synthesizes a complementary strand of RNA using an RNA template.

enzyme. A single phage tactic resolves both. The chro-
mosomes-to-be   are   first
packaged into new capsids
as ssRNA, and then the
complementary strands are
added by the RdRPs there.
Thus, Shy demonstrates that
an entire phage replication
cycle can be completed without risking endonuclease
attack, but this ploy comes
at a cost. It constrains the
phage's options for both the
regulation of transcription
and virion architecture. Shy
is one of the very few phages
who have chosen this route.

It is also possible to dodge
the waiting RNase III an-
other way. RNase III strikes
dsRNA; Minimalist[26] dodges by using ssRNA for its
chromosome. However, the
situation is more complex
because using ssRNA comes

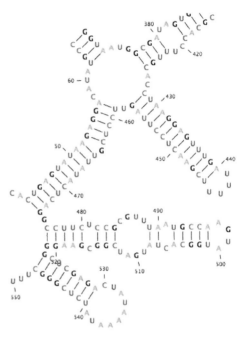

Figure 35: RNA phage dodges RNase
III. Minimalist uses intramolecular base
pairing skillfully to provide three-dimensional chromosome structure without nuclease risk. As exemplified in this
small segment of its chromosome, none
of the regions of perfect base pairing exceed 17 bp. RNA secondary structure was
generated from GenBank accession number AB971354.1 by Heather Maughan,
Ronin Institute, using CLC Main Workbench version 6.7.1.

with a price. A strand of ssRNA might be quite floppy and might assume random configurations in the cytoplasm. Minimalist needs the
benefits of a specific three-dimensional RNA structure for its mode
of production (see "The Ingenious Minimalist" on page 92). By being ingenious, it can have such a structure without succumbing to endonuclease attack. Almost its entire RNA chromosome is engaged in
intramolecular base pairing that provides dependable structures, such
as stem-loops, at specific points in the chromosome. Even though almost every nucleotide is base paired, RNase III can not attack. In order

---

[26] Minimalist: phage Qβ, a levivirus that infects *E. coli*.

to recognize and cleave dsRNA, RNase III requires a perfect double-stranded helix of at least 17 base pairs. None of Minimalist's stem-loops meet that criterion because the long stems include short bubble regions that exempt the region from nuclease attack (see Figure 35).

These are only some of the pitfalls awaiting an incoming phage chromosome on arrival. Some of these defenses are extremely effective by our military standards, but none are absolutely impenetrable to the phage. To tease apart these tactics, lab researchers use simple one phage–one host models; outside the lab, in any natural ecosystem, every bacterium has multiple defense tactics at hand to fend off not only many different phages but also other invaders (see PIC). Shrewd Bacteria attempt to avoid these intracellular phage confrontations altogether, to not let the wolf enter in the first place. They hide or lock the door, but have to change the locks frequently (see "Dodges" on page 195). The ascendance of even the best defense is inevitably temporary, and likewise the effectiveness of any phage counter tactic is provisional. If a phage chromosome makes it past all the cell's defenses intact, the phage swings into action to adeptly convert the bacterium, which had been previously intent on growing and dividing, into an efficient virion factory.

Further Reading

Bair, CL, D Rifat, LW Black. 2007. Exclusion of glucosyl-hydroxymethylcytosine DNA containing bacteriophages is overcome by the injected protein inhibitor IPI*. J Mol Biol 366:779-789.

Barrangou, R, C Fremaux, H Deveau, M Richards, P Boyaval, S Moineau, DA Romero, P Horvath. 2007. CRISPR provides acquired resistance against viruses in prokaryotes. Science 315:1709-1712.

Bondy-Denomy, J, B Garcia, S Strum, M Du, MF Rollins, Y Hidalgo-Reyes, B Wiedenheft, KL Maxwell, AR Davidson. 2015. Multiple mechanisms for CRISPR-Cas inhibition by anti-CRISPR proteins. Nature 526:136-139.

Dryden, DTF. 2006. DNA mimicry by proteins and the control of enzymatic activity on DNA. Trends Biotechnol 24:378-382.

Dy, RL, C Richter, GP Salmond, PC Fineran. 2014. Remarkable mechanisms in microbes to resist viral infections. Annu Rev Virol 1:307-331.

Hinton, DM. 2010. Transcriptional control in the prereplicative phase of T4 development. Virol J 7:289.

Hynes, AP, M Villion, S Moineau. 2014. Adaptation in bacterial CRISPR-Cas immunity can be driven by defective phages. Nat Commun 5:4399.

Labrie, SJ, JE Samson, S Moineau. 2010. Bacteriophage resistance mechanisms. Nat Rev Microbiol 8:317-327.

Olsthoorn, R, J Duin. 2011. Bacteriophages with ssRNA. eLS. http://bit.ly/2jth3Mf

Richter, C, JT Chang, PC Fineran. 2012. Function and regulation of clustered regularly interspaced short palindromic repeats (CRISPR)/CRISPR associated (Cas) systems. Viruses 4:2291-2311.

Samson, JE, AH Magadán, M Sabri, S Moineau. 2013. Revenge of the phages: Defeating bacterial defences. Nat Rev Microbiol 11:675-687.

Tock, MR, DTF Dryden. 2005. The biology of restriction and anti-restriction. Curr Opin Microbiol 8:466-472.

Westra, ER, DC Swarts, RH Staals, MM Jore, SJ Brouns, J van der Oost. 2012. The CRISPRs, they are a-changin': How prokaryotes generate adaptive immunity. Ann Rev Genet 46:311-339.

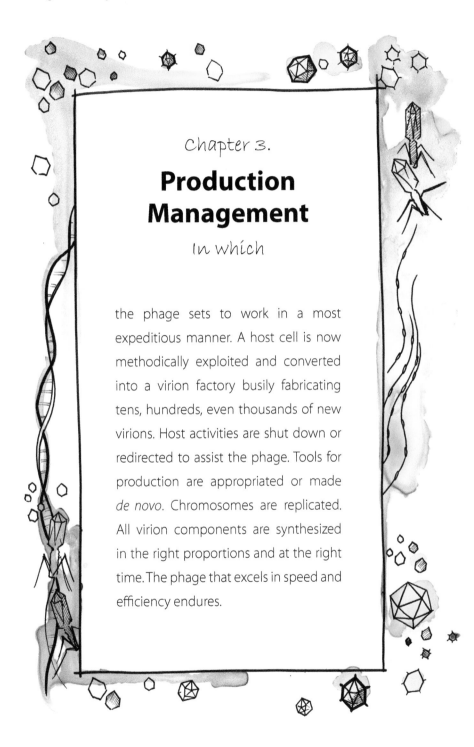

Chapter 3.

Production Management

In which

the phage sets to work in a most expeditious manner. A host cell is now methodically exploited and converted into a virion factory busily fabricating tens, hundreds, even thousands of new virions. Host activities are shut down or redirected to assist the phage. Tools for production are appropriated or made *de novo*. Chromosomes are replicated. All virion components are synthesized in the right proportions and at the right time. The phage that excels in speed and efficiency endures.

Creativity is key to productivity and prosperity.
Ifeanyi Enoch Onuoha

We have done so much, for so long, with so little,
we are now qualified to do anything with nothing.
Konstantin Jireček

Manufacturing takes place in very large facilities.
If you want to build a computer chip, you need a
giant semiconductor fabrication facility. But nature
can grow complex molecular machines using
nothing more than a plant.[1]
Ralph Merkle

When you make machines that are capable of
obeying instructions slavishly, and among those
instructions are "duplicate me" instructions, then of
course the system is wide open to exploitation by
parasites.
Richard Dawkins

---
[1] ...or phage and its host!

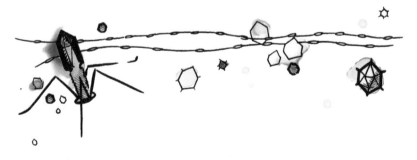

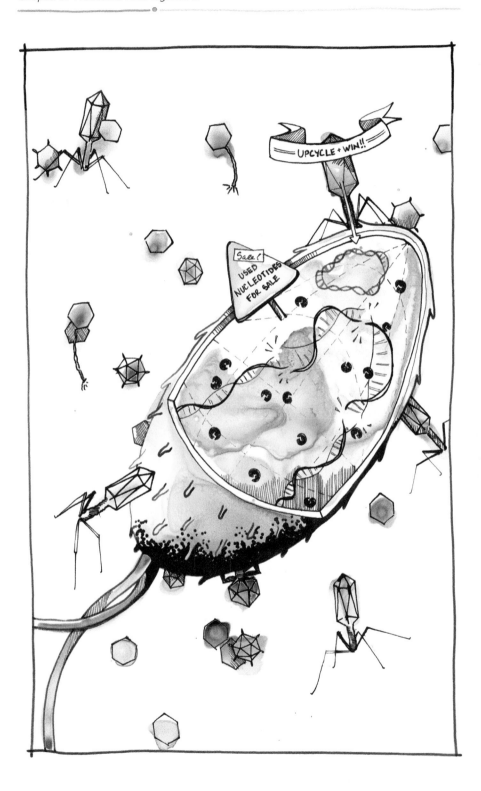

A phage chromosome arrives in an organized beehive of cellular enterprise. Raw materials are being selectively imported and wastes deliberately exported. All metabolic activities are coordinated and optimized for the current conditions. The problem for the phage is that all of this bustle is organized for maximum growth of the cell, with the goal of ultimately producing two cells from one. The goal of the phage is to redirect the whole enterprise toward virion production, while simultaneously protecting itself from any further defensive maneuvers by a rebellious host.

Each prokaryotic cell is a valuable resource sought after by numerous phages. Competition is fierce. The phage that survives in the long run is the one that produces the most progeny in the least time from whatever resources are at hand (see PIC). A crude takeover simply won't do. Takeover timing must be precise, efficiency maximized. Innovations that accelerate replication or use resources more efficiently are favored by natural selection. Each time a phage sharpens its management skills, it produces more progeny and its lineage will, as a result, become more abundant within the population. Phages have been honing their strategies for billions of years. They are ancient, but not primitive. Today's phages are highly evolved, strikingly skillful, and continually adapting.

Management Basics

Throughout an infection, the phage manipulates the metabolism of the virocell to support production of progeny virions. The tactics employed are as diverse as the phages themselves. Is there an optimal strategy? For example, should the phage encode only a few, absolutely essential, phage-specific proteins in its chromosome and rely on the host to supply everything else? This would be a sensible policy. However, a phage might fare better if it brought along its own genes for proteins to carry out some key cellular functions. These genes could be optimized to best serve the phage's ambitions. Moreover, this eliminates the danger that the host might modify or eliminate a protein that the phage requires. A similar question concerns the optimal treatment of the host cell. Should the phage destroy the host chromosome, block cellular growth, and scrupulously disrupt all activities that are not essential for phage reproduction? A crippled virocell is less able to respond to environmental changes and gradually the resources available to the phage dwindle. It might be wiser for a phage to allow the virocell to continue to grow, even to divide, while the phage discreetly siphons off the goods and services it needs. It seems that some phages pursue every strategy we can imagine, and also some that we never imagined.

What does phage production require? Every progeny virion that comes off the assembly line contains proteins and a chromosome (DNA or RNA); a few phages also add a lipid membrane. Synthesis of phage proteins requires a supply of amino acids, the translation machinery (ribosomes, tRNAs, and associated factors), and energy to fuel the process. The larger and the more complex the virion, the more proteins invested in its construction, and thus the greater the demand for both amino acids and energy. Likewise, chromosome replication requires an abundance of nucleotides, the machinery for chromosome replication, and energy. The phage depends on its host cell to supply most of these essentials. All energy production is the work of cellular metabolism. In choosing its strategy here the phage walks a thin line. Phage takeover tactics may knock out host protein synthesis in order to block production of new antiphage weapons, but this may also cripple host energy production and thereby cripple phage replication (see "Photosynthetic Phages" on page 102).

For the synthesis of their needed macromolecules,[2] phages usually rely on the machinery present in every cell. Every prokaryote maintains pools of amino acids and nucleotides for its own use, which the phage promptly commandeers. Often this supply is not adequate to satisfy the extraordinary demands of phage replication. To bridge the shortfall, some phages dismantle their host's DNA, RNA, and proteins, then upcycle the components into new virions. Even this may still fall short. Phage chromosome replication places unprecedented demands on the host's supply of phosphate, as copious phosphate is consumed by DNA and RNA synthesis. In a phosphate-poor environment such

---

[2] macromolecule: a molecule containing a very large number of atoms, often assembled by linking smaller molecules termed building blocks. For example, proteins, nucleic acids, carbohydrates, and lipids.

as the open ocean, the host may have been barely scraping by before the phage arrived, barely meeting its own needs. Many marine phages take the plight of their hosts into account. These phages carry genes that help the virocell to scavenge the extra phosphate needed to support phage DNA synthesis. This scenario is but one example of how phages may appear to assist the virocell, while in actuality they are looking after their own interests.

Phage replication is an efficient assembly line operation. Whether the phage genes are few or many, their activity is precisely regulated. Timing is critical: when in the orderly course of an infection is each gene product required? Some proteins are "early proteins" that are needed immediately upon arrival, while others are not required until later, perhaps after phage chromosome replication is well underway. The phage schedules its parts production and manages its inventory to maximize efficiency, avoid shortages, and minimize waste. Production of each protein is matched to market demand – only a few copies of some but hundreds or thousands of others. These strategies are sophisticated, ingenious, and time-tested in a vast marketplace. Comparing Minimalist, Lander, and Skinny,[3] all of which infect the same bacterial species, demonstrates how very different these strategies can be.

The Ingenious Minimalist

Can a "simple" phage possibly execute a sophisticated replication program with a mere three genes? Minimalist (Qβ) says "Yes!" Moreover, this three-gene marvel competes successfully with phages such as Lander with its close to 300 genes for a share of the available *E. coli* hosts. Three genes doesn't sound like enough to carry out all the steps in a lytic life cycle: host recognition and entry, evasion of host anti-phage defenses, chromosome replication, virion construction, and the timely release of progeny from the virocell. In addition to that list, Minimalist must also encode the machinery needed to replicate its single-stranded RNA (ssRNA) chromosome because this is something no host cell knows how to do.

How does Minimalist spend its three gene budget? One gene, its replicase gene, is dedicated to replication of its ssRNA chromosome. Actu-

---

[3] Skinny: enterobacteria phage f1, an inovirus that infects *E. coli*.

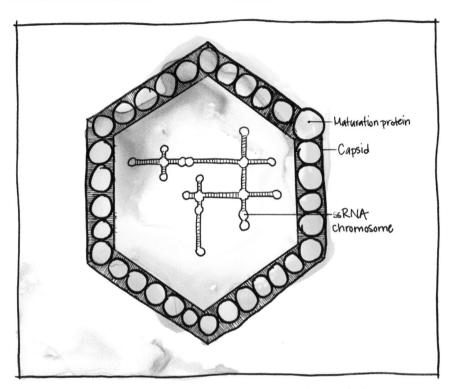

Figure 36: A simple virion floor plan. All that Minimalist needs for its economical virion is one maturation protein, 180 major capsid proteins, and one ssRNA chromosome.

ally it takes four proteins to form the functional RNA replicase holoenzyme: the phage-encoded replicase and three others borrowed from the host. Those three are cellular proteins that are normally engaged in translation, so Minimalist can rely on the virocell to dependably provide them. A little judicious thievery can make a gene go farther.

Another one of Minimalist's genes encodes the major capsid protein, the main constituent of its small (26 nm) icosahedral capsid (see Figure 36). The protein product of the third gene, the maturation protein, is a multi-tasking workhorse. Only one copy is included in each mature virion, but it is essential there. Without it, the RNA inside would be vulnerable to digestion by environmental RNases. It also is the virion component that recognizes and adsorbs to the host, after which it escorts the chromosome into the cell. Lastly, this same protein determines when it is time to lyse the virocell and then does the deed itself (see "How to Lyse Your Virocell Using Only One Gene" on page 169).

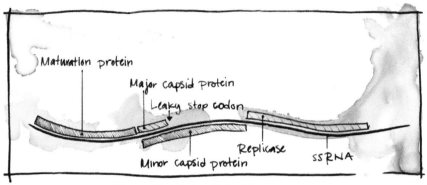

Figure 37: Minimalist's Chromosome

Minimalist's chromosome arrives in the host as positive-sense ssRNA. This means that the chromosome itself can serve as mRNA and is ready to be translated by the cell's ribosomes immediately upon arrival. However, it is a polycistronic mRNA that contains all three genes in each molecule (see Figure 37). This presents a challenge: how to match the relative numbers of each protein type made to the needs of the phage at each stage during the infection. Cells with their DNA chromosomes can use a separate mRNA for each gene or for small clusters of contiguous genes whose products are needed together and in equal numbers. Individual genes or clusters can then be silenced or activated by regulating the amount of mRNA transcribed from them. This tactic – transcriptional regulation – is not an option for Minimalist because every one of its mRNAs contains all three genes. The only option available, then, is to regulate the rate at which each gene is translated into protein. For this, cells can afford to employ translation regulator proteins, but Minimalist can't afford a separate gene to encode even one regulatory protein. Instead it relies on an elegant solution that is inherent in the base sequence of its ssRNA chromosome.

DNA or RNA?

DNA is boring. Granted, DNA's double-helix and chemical stability makes it well suited for archiving genetic information. DNA is what comes to mind when we think about genes and genomes. But RNA can also serve as the stuff of chromosomes, encoding and replicating genetic information in much the same way as DNA. It is thought by some that RNA comprised the chromosomes of the earliest life forms, and only later in evolution was the task of archiving genetic information handed off to DNA. Information storage is DNA's primary service, whereas RNA carries out many diverse cellular functions. RNA is full of surprises. It constitutes much of the machinery for protein translation (e.g., ribosomes, tRNAs, mRNA), functions as ribozymes (within ribosomes and elsewhere), and as riboswitches that regulate protein translation in response to environmental conditions. This functional versatility stems from RNA's structural diversity. dsDNA assumes only one basic configuration, that being a double helix in which two complementary strands are joined by hydrogen bonds at each base pair. dsRNA can form a similar helix. However, the RNAs that carry out many cellular functions are structurally creative ssRNAs that form intricate, highly specific, three-dimensional structures, such as the tRNAs with their cloverleaf structure (see "Nucleic Acid Structure" on page 56). Key to the more complex structures are short regions of inflexible double helix. As usual, this helical structure requires base pairing between two complementary strands, but in this case the "strands" are two regions within the same RNA molecule. A typical ssRNA molecule, including Minimalist's chromosome, does not mimic a tangle of overcooked vermicelli, but instead assumes a specific three-dimensional configuration. The chromosome loops and folds back on itself neatly and reliably to maximize the number of stabilizing hydrogen bonds.

Here Minimalist exploits the potential benefits of its ssRNA chromosome, aka mRNA, to the max by using the three-dimensional configuration of its chromosome to regulate translation. Almost its entire chromosome is reliably engaged in stem-loop structures in which double-stranded regions form the helical stems and short, single-stranded zones comprise the loops at the ends. Schematic "road-kill" diagrams (see Figure 38) are two-dimensional representations of the secondary structure. While these do convey the amount of intramolecular base pairing, they also mislead. In your imagination transform this flatland

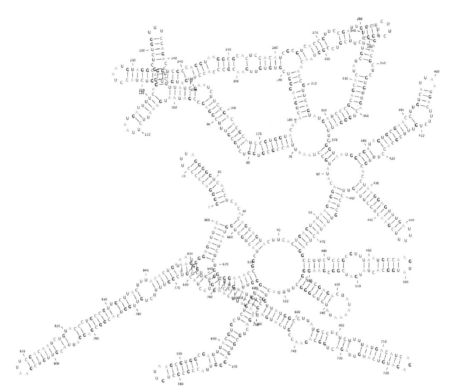

Figure 38: Minimalist as road kill. A "road kill" diagram of the first 20% (860 nts) of Minimalist's ssRNA chromosome. RNA secondary structure was generated from GenBank accession number AB971354.1 by Heather Maughan, Ronin Institute, using CLC Main Workbench version 6.7.1.

vision into a precisely-folded, complex, three-dimensional structure in which some regions are accessible, but others are buried. This property enables Minimalist to regulate the synthesis of each protein—both when and how many copies are made.

To make a protein, a ribosome must attach at its specific binding site on the mRNA. It locates the nearby start codon and initiates a new protein chain there. Then it chugs along the mRNA, adding amino acids one by one to the growing protein chain as it goes. This process continues until the ribosome encounters a stop codon, at which point the completed protein is released and the ribosome falls off the mRNA. If the ribosome binding site at the start of a gene is inaccessible due to the folded configuration of the mRNA, that protein will be translated rarely.

This interference is all the opportunity that Minimalist required. By judicious tweaking of the base sequence of its chromosome, combined with a few other sly phage tricks, Minimalist regulates the translation of each of its proteins independently. Upon arrival in a host cell, Minimalist's ssRNA chromosome folds and base pairs to assume a stable configuration that allows access to only one ribosome binding site, that being the one for the capsid protein. Production of the capsid proteins begins immediately and proceeds at maximum rate throughout the infection.

That straight-ahead tactic is fine for the capsid protein, but not for the others. For the replicase, only a few copies are sufficient since one replicase can catalyze the synthesis of many new chromosomes. The replicase is wanted immediately so that chromosome replication can get underway, but initially ribosomes are blocked from translating this gene. However, as ribosomes busily synthesize coat proteins, this activity disrupts the folded ssRNA structure enough to allow other ribosomes to now access their binding site at the start of the replicase gene. Before long, sufficient replicase has been made. Now it is time to shut down replicase synthesis to conserve resources, without impeding major capsid protein production. Minimalist needs a way to assess when enough replicase has been made, as well as a mechanism to turn its synthesis off. It uses the ongoing accumulation of capsid proteins to do both. As more and more capsid proteins collect, more of them form dimers. These dimers have a tendency to bind to the nearby start of the replicase gene, thus hiding it from the ribosomes and shutting down replicase synthesis.

What about the third gene, the one encoding the maturation protein? Since only one copy is used in constructing each new virion, the ideal scheme would be to make only one maturation protein for each new RNA chromosome. Minimalist has just such a scheme, one that makes use of the directionality of both RNA and protein synthesis. This tactic is built on two interwoven properties of Minimalist's replicase. This replicase can use either positive-sense or negative-sense ssRNA, but not dsRNA, as its template. As it works, it also prevents the formation of hydrogen bonds between the template and the newly minted product strand. During each infection, the first task for replicase is to synthesize a negative-sense ssRNA complementary to the infecting, positive-sense chromosome. With both strands now available, replicase can proceed to synthesize more and more positive-sense chromosomes and negative-sense templates.

During the synthesis of any new DNA or RNA strand, including new positive-sense ssRNA chromosomes, the 5′ end is made first. Nucleotides are then quickly added sequentially to extend the strand all the way to the 3′ end. Since Minimalist's chromosome is also the mRNA, the gene for the maturation protein is the one nearest to the 5′ end of the mRNA, and this gene is the first to be synthesized. Protein translation is also directional. Ribosomes bind at the 5′ end of a gene and move toward the 3′ end, translating as they go. Therefore, their binding site lies in the first region of the mRNA to be synthesized. A ribosome can hop onto the mRNA and start protein translation immediately, even before the entire mRNA (i.e., the entire chromosome) has been synthesized. Minimalist takes advantage of this. As each new chromosome is being made, a ribosome can slip in and initiate translation of a maturation protein. Meanwhile the replicase moves on down the template ssRNA ahead of it to synthesize the rest of the chromosome/mRNA. Very quickly the new ssRNA strand is long enough to fold back on itself and form the three-dimensional structure that will block further ribosome access. This effectively limits maturation protein synthesis to about one copy per chromosome. Occasionally a second ribosome will bind before access disappears, so a few extra copies do get made. This is an advantageous feature, not a defect. Those extra copies accumulate and ultimately lyse the virocell (see "How to Lyse Your Virocell Using Only One Gene" on page 169).

One other issue remains: capsid protein production. Assembly of each virion calls for 180 copies of the major capsid protein. As mentioned above, abundant copies are produced by continuous rapid translation throughout the infection. However, each virion also requires twelve copies of a minor capsid protein, a variant that is a bit longer. Here Minimalist uses a common phage trick to get two different proteins for not much more than the chromosomal cost of one. That trick is to end the gene for the shorter protein with a "leaky" stop codon. During capsid protein synthesis, the ribosome usually stops at this codon on the mRNA and terminates the new protein chain. Occasionally, the ribosome ignores the stop sign and continues translating further along the mRNA to extend the protein. It goes on adding more amino acids until it reaches the next stop codon. Thus, both proteins are identical up to that leaky stop codon while one of the two extends further. The ratio of the quantities made depends on the leakiness of the stop codon. For Minimalist's capsid proteins, the ribosome "reads through" the leaky stop codon about 5% of the time to yield enough copies of the minor capsid protein for virion assembly – no other regulatory controls required.

Minimalist is not unique in employing these particular mechanisms to increase its production efficiency. The same approaches are used by phages with far larger genomic budgets. Minimalist won a place in the spotlight here because it is the efficient phage par excellence. Equipped with these stratagems, it is capable of producing thousands of virions from one host cell. The yield is greater when the host is living in plush conditions, as when supplied with abundant food and a hospitable environment in the lab. Then the Bacteria are larger, contain more ribosomes, and grow faster. Given such luxuries, Minimalist completes its infection cycle in an hour, producing more than 10,000 progeny. Unlike some phages, it does not shut down host macromolecule synthesis, but it does divert about half of the host's metabolic activity to the production of phage RNAs and proteins. Consider what a massive undertaking it is for the cell to supply the building blocks and energy to make perhaps 5000 new phage chromosomes even when each is only a small ssRNA molecule. Each Minimalist chromosome contains 4,217 nucleotides ($\sim 4 \times 10^3$ nts). Multiplying that by 5,000 (5×10^3) progeny gives you a total of twenty million (20×10^6). The genome of a typical lab strain of E. coli contains about four million ($\sim 4 \times 10^6$) base pairs or about eight million (8×10^6) nucleotides. Each time the cell divides to form two daughter cells, it needs to synthesize only eight million nucleotides versus the twenty million it makes for the phage chromosomes. In addition, although minor compared to the cell's own protein complement, it has supported the synthesis of at least a million phage proteins, each with hundreds of amino acids. Its reward for services rendered? Lysis by the phage.

The Lander Method

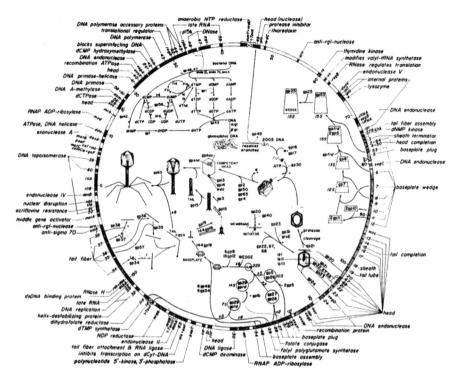

Figure 39: The Lander method. Lander has 280 genes available to provide tools to carry out each step in an infection, counter host defenses, compete with other phages, exploit its host, and assemble a flamboyant virion. Here representative genes are identified around the perimeter of Lander's circular chromosome. The diagrams inside the circle relate some genes to its virion assembly pathway (see "A Lipid Supplement" on page 144) and others to the synthesis of its modified DNA (see "Survival on Arrival" on page 61). Courtesy of Elizabeth Kutter and Burton Guttman, Evergreen State College, Olympia.

There is more than one road to success. Both Minimalist (Qβ) and Lander (T4) effectively convert *E. coli* into short-term virion factories, but their methods could hardly be more different. Minimalist does it with a ssRNA genome that contains three genes; Lander undertakes the same challenge equipped with a dsDNA chromosome with 280 genes (see Figure 39). If Minimalist can successfully replicate in *E. coli* with but three genes, why does Lander have so many? What functions do they serve?

Of those 280 genes, only 62 are essential for Lander to replicate when provided with a culture of *E. coli* in the lab. These 62 genes, being longer than many of the others, account for almost half of the genome. They include most of the 40 genes that encode virion structural proteins as well as the seven proteins that carry out cell lysis. Others redirect cellular metabolism, replicate Lander's "sweet" DNA (see "The Saga of Sweet Lander" on page 73), or assist with virion construction. We don't know what almost half of Lander's genes do. This lack of knowledge is not surprising given how we attempt to determine gene function. Commonly we delete or inactivate a gene, then ask whether Lander can still grow normally on a standard strain of *E. coli* in the lab. If the phage is unable to replicate its chromosome, assemble a normal capsid, or lyse the host, or if it displays some other observable glitch, we can investigate further and often track down the exact role of that gene. However, if the phage still replicates normally, we're left in the dark.

A laboratory culture is an unnatural environment. Here nutrients are abundant and there is no competition from other types of Bacteria or phages. These simple phage-host model systems have been exceedingly useful in deciphering the intricacies of phage replication, but they exclude the many convoluted interactions that a phage experiences in natural environments (see PIC). There are always attempts by other phages, both related and unrelated, to gain possession of the same host. Even after you have staked your claim and initiated an infection, other phages may still attempt to hijack "your" virocell. To counter such moves, many phages carry genes that act to block subsequent infection by another phage.[4] And, of course, the pirates may carry genes to overcome those blocks. Similarly, hosts are perpetually modifying or augmenting their defenses. A phage gene that is not required to infect one host strain may be necessary to survive the antiphage defenses of another.

In many environments, a phage's intended host is already coping with nutrient shortages or out-and-out starvation. Phage replication adds additional demands for energy and nutrients. In addition, the takeover strategy of some phages, such as Lander, includes immediate shutdown of host protein synthesis upon arrival, further exacerbating the situa-

---

[4] superinfection exclusion: various processes whereby a phage within a virocell actively blocks infection of that cell by another related phage.

tion. Since a phage's reproductive success depends on virocell metabolism, it pays for a phage to lend a hand and help the virocell to sustain despite adverse external conditions and phage-generated disruptions.

For example, the synthesis of many copies of a large phage chromosome such as Lander's requires abundant phosphate, a nutrient that is scarce in some environments. A phage carrying a gene that boosts phosphate scavenging can make more phosphate available for the virocell and thus allow production of more phage progeny. This capability gives the phage a competitive advantage over a relative who lacks the gene. Likewise, a phage might carry genes that enable the virocell to detoxify an environmental poison or degrade an antibiotic, thus keeping the cell alive long enough for the phage to replicate. Phages are now known to carry a diverse array of genes that contribute to virocell metabolism, genes collectively referred to as auxiliary metabolic genes (AMGs).[5] Several of Lander's AMGs are dedicated to one of its anti-restriction defenses: the replacement of every cytosine (C) in its DNA with hydroxymethyl cytosine (HMC) (see "The Saga of Sweet Lander" on page 73). These AMGs include genes for the enzymes that divert cytosine from the DNA synthesis pathway and convert it into HMC, then one for another enzyme to add a glucose to each HMC, and one more to encode an HMC-tolerant DNA polymerase.

Photosynthetic Phages

The most abundant photosynthetic microbes in the oceans are two small cyanobacteria, *Synechococcus* and *Prochlorococcus*. The smaller *Prochlorococcus* is the more numerous, with a global population of $\sim 10^{27}$ cells, but *Synechococcus* is not far behind with $\sim 10^{26}$ cells. Together they account for a major share of marine photosynthetic activity, and thus are major players in global food webs and carbon cycling. These Bacteria also support large numbers of cyanophages, including some of Lander's relatives. Like all phages, these cyanophages rely on the virocell to provide abundant energy to fuel their replication. This means that cyanophages are dependent on virocell photosynthesis

---

[5] auxiliary metabolic gene: (AMG) phage genes that function in cellular metabolism. Initially they were thought to be the exclusive property of cells and only incidentally associated with phage.

throughout their infection cycle. Many, like Lander, knock out host protein synthesis as part of their cellular takeover strategy. This tactic creates a catch-22.

Capturing the energy of sunlight is a suicidal activity for the protein heterodimer located in the reaction center of the cell's light harvesting apparatus. Both proteins suffer light-induced damage. One of them (D1) turns over particularly rapidly and must be continually replaced by newly synthesized copies in order for photosynthesis to continue. Turn off synthesis of host proteins, and soon photosynthesis declines, to the detriment of the phage's replication goal. You are, no doubt, confident at this point that these cyanophages have a solution, perhaps even more than one. Indeed they do. One tactic used by the majority of cyanophages is to encode their own gene for protein D1. When synthesis of the host's D1 protein falls off, active transcription of the phage gene ensures that enough copies of the phage D1 protein are made to keep photosynthesis operational to the end. Many cyanophages also carry genes for other components needed to power the virocell by the sun and to reroute the output to preferentially support phage replication.

About 50 of Lander's genes encode small "monkey-wrench" proteins that facilitate host takeover by interfering with host metabolism. When Lander first arrives in the cell, the host's 20,000 or so RNA polymerases (RNAPs) are busily transcribing host genes, and those transcripts are being translated into host proteins. Lander's first job is to put that machinery to work synthesizing phage proteins instead – and not all phage proteins, only the ones needed early in the infection. What determines which genes are transcribed? In part this is the interplay between the promoter associated with each gene and a subunit of the RNAP holoenzyme: the sigma factor.[6] Genes with highly similar promoters will be recognized and preferentially transcribed by RNAPs that contain a particular sigma factor. Substitute a different sigma

---

[6] sigma (σ) factor: the subunit of the RNAP holoenzyme that recognizes and facilitates binding of RNAP to specific promoters, thereby initiating transcription of the downstream gene(s).

factor, and you set the RNAPs to work transcribing a different group of genes. The bacterium uses this tactic to regulate transcription of a group of selected genes in concert, even when those genes are scattered across its chromosome. If, for instance, food suddenly becomes scarce, it can quickly redirect transcription to the numerous genes that help it cope with starvation no matter where in the chromosome those genes are located.

This is indeed a useful tactic for the cell, but it is also open to exploitation by a phage. Lander's exploitation begins immediately upon arrival. Along with its chromosome, it delivers a protein[7] that modifies the host's RNAPs so as to redirect them to transcribe the genes for Lander's early proteins. Some of those early proteins made within the first two minutes continue the host manipulation. One[8] halts currently ongoing transcription of host DNA, while two others prevent further transcription by cleaving the DNA. Yet others act at five or seven minutes post-infection to modify the sigma factor to direct the RNAPs to the genes needed first for phage DNA replication, and then to those needed for assembly of progeny virions. In short, many of Lander's 280 genes earn their place in the genome by fine-tuning production efficiency.

And the Winner Is?

Such different infection strategies prompt one to ask which phage is more successful, Minimalist or Lander. Without some specific criterion of success, the question is meaningless. In one sense they are equally successful: both are winners in the highly competitive, ongoing evolutionary game. They both have endured. On the other hand, Lander is a member of a large clan, with at least five major subgroups. Its close relatives are abundant, widespread, and diverse. Moreover, they are members of the order Caudovirales, the tailed phages. The virions of these phages all have a tail attached to one vertex of their icosahedral capsid. By 2007, the virions of 5,568 phages had been observed under the electron microscope. Of these, 96% were tailed icosahedral capsids. The remaining 4% include tailless icosahedra such as Minimalist, as

---

[7] Alt protein: an internal protein of Lander, about 40 copies of which are delivered during infection along with its DNA chromosome.

[8] Alc protein: one of the proteins made by Lander during the first two minutes after arrival. It acts as a transcription terminator that halts transcription at specific C-containing sequences, thus selectively interfering with transcription of host genes.

well as a variety of unexpected forms. Thus, it seems that a tailed virion gives a phage an advantage, but phages have experimented with diverse architectures, some of which work quite well in particular niches.

Under favorable lab conditions, Minimalist produces 10,000 or more progeny virions during an infection, Lander perhaps 200–300. The numbers for both are much lower in the environment. Despite this numerical advantage, Minimalist has not nabbed all the *E. coli* cells, has not thwarted Lander's virions. Why not? Less than half – perhaps only 10% – of Minimalist's virions are infectious when they exit the virocell. One could speculate that this may be a result of errors in assembly or in replication of their RNA chromosome. Or it might reflect a trade-off for some other advantage gained. After exiting, the probability of chromosomal death is high for Minimalist, higher than for Lander, because ssRNA is more vulnerable than dsDNA to hazards such as UV irradiation. If Lander's dsDNA chromosome is injured by UV in transit, that damage is sometimes repaired after arrival by the host's DNA repair machinery – a courtesy not extended to Minimalist's RNA chromosome. Moreover, even though Minimalist and Lander both infect the same bacterial species, the probability of making a productive contact with an *E. coli* cell is lower for a Minimalist virion. To recognize a potential host and launch an infection, these virions must contact a specific cell appendage, a sex pilus,[9] and from there make their way into the cell (see "Bacterial Conjugation" on page 206). Only some *E. coli* strains make sex pili, and even they do this only part of the time. The more virions Minimalist produces during an infection, the more likely it is that at least one infectious virion will bump into a sex pilus before it expires.

Each phage production strategy is intertwined with every step in its life cycle and is impacted by a phage's choice of chromosome type. For Minimalist, production of 10,000 virions from a brief infection requires minimizing the consumption of cellular resources per virion produced. This, in turn, necessitates a smaller capsid and a smaller chromosome, and thus fewer genes dedicated to countering host defenses, fine-tuning the accuracy of chromosome replication, competing with co-infecting phages, or other activities.

---

[9] sex pilus: a pilus that is essential for conjugation and that is typically encoded by a conjugation plasmid.

So Wasteful!

Despite their differences in strategy, both Minimalist and Lander consider the highest and best use of a virocell to be immediate, intense exploitation, promptly terminated by lysis. This is the most-studied phage strategy, for both historical and practical reasons. However, it is not the only phage way, nor even the most popular one. This pattern of infection treats the virocell as a disposable resource. Rapid exploitation of the virocell does yield the highest short-term profit, i.e., the most virions in the least amount of time. But Skinny (Ff), who also infects *E. coli*, regards it as wasteful to blow up the factory just to ship your virion products out the door. In each infection cycle, time is lost setting up a new production line. There is also the downtime wasted while searching for the next host. Skinny, instead, spends the first ten minutes after it arrives in a cell setting up shop. Within 30 minutes it is in stride, steadily producing virions at a rate that can continue indefinitely. Its virions quietly slip out through the cell membrane, a thousand or so each cell generation (see "A Different Story" on page 160). The released filamentous virions even benefit the virocell by contributing to biofilm[10] formation. However, although the virocell lives on, its growth is markedly slowed, which gives its noninfected competitors an advantage. Utilizing extrusion for release of the progeny virions also exacts a price on the phage. It puts major constraints on virion architecture (a simple, flexible rod; see "Live and Let Live...and Exploit" on page 149), chromosome type (circular ssDNA), and genome size (only 10 genes for Skinny). Today, Skinny's clan constitutes only a minor share of the virosphere.

There is yet another strategy that not only makes longer term use of the virocell but is also widely employed—perhaps the majority choice. It delivers the benefits of long-term residence in the virocell combined with the rapid virion production of lytic infection. The drawback? Delayed gratification. Upon arrival, instead of going directly into production, the phage settles into cellular life, often by integrating its chromo-

---

[10] biofilm: a structured community of microbes that adhere to each other and also typically to a surface. Often they live embedded within a matrix of secreted extracellular polysaccharides combined with proteins and DNA. The matrix provides protection while maintaining the structure that facilitates cooperative interchanges between cells.

some into the host chromosome. Instead of producing 25, 100, or 1000 progeny rapidly, it makes do with a mere duplication each time the virocell divides. This strategy, lysogeny, is an intricate, co-evolved relationship, with benefits to both parties – a story that warrants its own chapter (see "Coalition" on page 241).

Behind every successful phage is a prokaryote cell whose energy, tools, and materials have been redirected to phage replication. The strategies employed by different phages vary, each one adapted to a particular phage-host pair. They range from the subtle diversion of resources to out-and-out takeover accompanied by immediate lethal damage. This behavior is what has earned the phage its designation as an intracellular parasite – a disparaging epithet indeed. Granted, from the perspective of the individual infected prokaryote, that label seems warranted. It is only at the community level that the essential services provided by the phages become dramatically apparent (see PIC). Each individual invader's mission is to efficiently convert the virocell into more phages. To that end, they supervise the timely production of all the components and machinery needed to efficiently feed their virion assembly line.

Further Reading

Beekwilder, J, R Nieuwenhuizen, R Poot, Jv Duin. 1996. Secondary structure model for the first three domains of Qβ RNA. Control of A-protein synthesis. J Mol Biol 256:8-19.

Clokie, MR, NH Mann. 2006. Marine cyanophages and light. Environ Microbiol 8:2074-2082.

Cvirkaite-Krupovic, V, R Carballido-López, P Tavares. 2015. Virus evolution toward limited dependence on nonessential functions of the host: The case of bacteriophage SPP1. J Virol 89:2875-2883.

Hendrix, RW. 2010. Recoding in bacteriophages. in *Recoding: Expansion of Decoding Rules Enriches Gene Expression*: Springer Science+Business Media. p. 249-258. http://bit.ly/2iKch9f

Mathews, C.K. 2001. Bacteriophage T4. eLS. http://bit.ly/2iv4xl2

Mai-Prochnow, A, JGK Hui, S Kjelleberg, J Rakonjac, D McDougald, SA Rice. 2015. Big things in small packages: the genetics of filamentous phage and effects on fitness of their host. FEMS Microbiol Rev 39:465-487.

Olsthoorn, R, J Duin. 2011. Bacteriophages with ssRNA. eLS. http://bit.ly/2iCzleG

Weissmann, C. 1974. The making of a phage. FEBS letters 40:S3-S9.

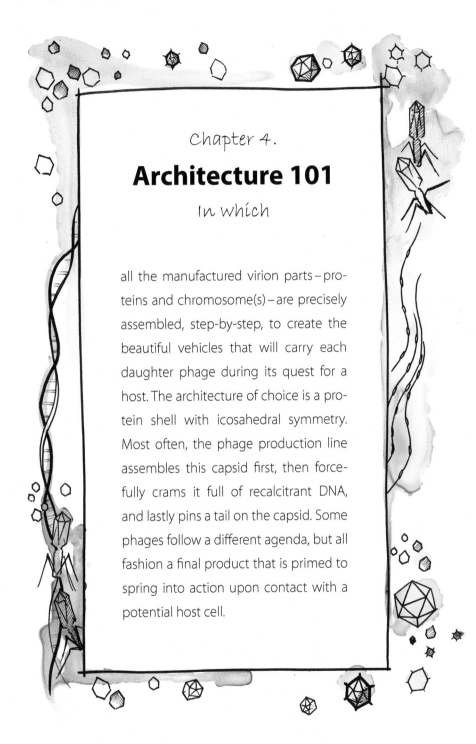

Chapter 4.

Architecture 101

In which

all the manufactured virion parts – proteins and chromosome(s) – are precisely assembled, step-by-step, to create the beautiful vehicles that will carry each daughter phage during its quest for a host. The architecture of choice is a protein shell with icosahedral symmetry. Most often, the phage production line assembles this capsid first, then forcefully crams it full of recalcitrant DNA, and lastly pins a tail on the capsid. Some phages follow a different agenda, but all fashion a final product that is primed to spring into action upon contact with a potential host cell.

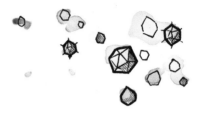

It is the pervading law of all things organic and inorganic, of all things physical and metaphysical, of all things human and all things superhuman, of all true manifestations of the head, of the heart, of the soul, that the life is recognizable in its expression, that form ever follows function. This is the law.

Louis H. Sullivan, *The Tall Office Building Artistically Considered,* **1896**

In seeking knowledge, day by day something is added. In following Tao, day by day something is dropped. Day by day you do less and less deliberately. Day by day you don't do more and more. You do less and less and don't do more and more, until everything happens spontaneously. Then you act without acting, and do without doing, and achieve without forcing. And nothing is done. And nothing is left undone.

Carol Deppe, *Tao Te Ching*

We've heard that tune: "simple, identical parts, identically assembled" fits the very definition of a Platonic solid! Because the part generates the whole, the virus does not need to "know" about dodecahedra or icosahedra…

Frank Wilczek, *A Beautiful Question*

If ever we are to attain a final theory in biology, we will surely, surely have to understand the commingling of self-organization and selection.

Stuart A. Kauffman

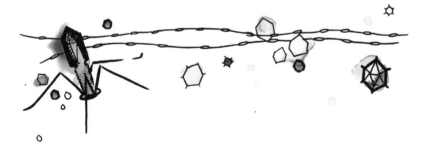

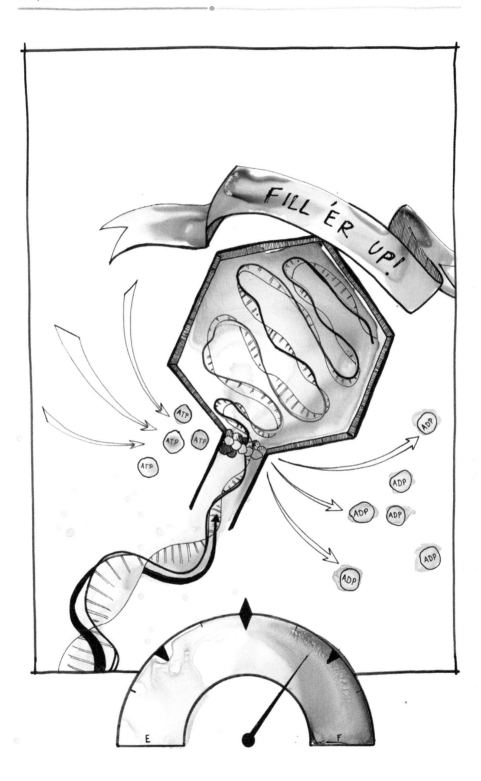

Virions are astonishingly beautiful. If we could see them with our own eyes, we would admire their architectural flair, their elegant functionality. A virion is the *sine qua non* of a phage, the feature that distinguishes phages from the assortment of mobile genetic elements that travel naked, such as plasmids, transposons,[1] and viroids.[2] Even this distinction blurs if examined too closely. The products of evolution do not fit into neat boxes. There are, for example, large transposons that encode capsid proteins. (Whether or not they construct capsids for extracellular transport remains an open question. Are these transposons on the way to becoming phages, or phages becoming transposons, or neither?)

Virions are also eminently functional, and viewed by some as the quintessential "nano-machine". Although that label is intended to express respect, to me it belittles them by putting them in the same category as the devices we fabricate and own. If they are "machines," they are the most exquisite of machines – multi-functional, requiring minimal material, capable of rapid self-assembly with an enviable level of quality control, and embodying an intrinsic architectural flair. Their form does follow their function, but this has not stifled phage creativity. Collectively they evolved several functional designs and explored variations on each theme. Yes, I am in awe. The diversity and perfection that we see today is the product of several billion years of innovation honed by relentless competition and selection.

From the moment a phage chromosome enters the cell until the first virion comes off the assembly line, the phage is in eclipse. There are no infectious particles anywhere to be found. Every phage generation creates its progeny virions from scratch, building them from newly made phage proteins and a newly replicated chromosome. This *de novo* synthesis of virions sets phages (and all other viruses) apart from all cellular life. When a cell reproduces, it divides into two daughter cells each of which inherits not only genes, but also part of the highly

---

[1] transposon: a common DNA element that moves from one location to another either within a chromosome or to another chromosome in the same cell.

[2] viroid: an infectious agent that infects plants and that consists of only a circular, single-stranded molecule of negative-sense RNA without a capsid.

structured cytoplasm of the mother cell. When a phage reproduces, all it brings to the task is the information encoded in its chromosome. It is the task of the virion to transport a dormant phage chromosome through intercellular space and deliver that information into a cell capable of supporting production of the next generation. En route a virion drifts through an environment teeming with hazards that threaten to damage the craft or cripple the chromosome inside. It collides with microbes, cellular debris, and all manner of flotsam, then rebounds, intact. Rebounds, that is, unless it has bumped into a potential host, in which case it must respond and quite literally spring into action. No metabolic energy can be generated en route, yet the virion must adsorb to the cell and deliver its cargo inside. Both stability and quick response are inherent in virion structure, and that structure, in turn, results from the protein sequences encoded in the genome. This process echoes the way the sequence of Minimalist's chromosome seemingly effortlessly regulates both the *what* and the *when* of phage protein synthesis (see "The Ingenious Minimalist" on page 92). Both are elegant solutions.

Architectural Basics

All virions, whatever their size or shape, are constructed from multiple copies of a few proteins. Moreover, those proteins are moderate in size for many of the same reasons that we build houses out of many small bricks rather than a few large ones. Phages with the smallest genomes make the simplest virions using the fewest different proteins. Often the capsid is constructed from multiple copies of just one major capsid protein, with perhaps another minor component that serves to recognize and bind a potential host cell (see "The Ingenious Minimalist" on page 92). Even phages with more genes adhere to the same basic capsid architecture, supplemented with a few additional proteins for useful spikes or "glue." At the other end of the spectrum is Lander who, among many others, assembles a decorated capsid and flaunts an elaborate tail. Each of Lander's virions contains multiple copies of about 40 different structural proteins.

With an eye to the cost of production, a phage architect also takes into account the total number of protein molecules invested in each virion. *Ceteris paribus*, a capsid geometry that requires fewer protein molecules to transport the same chromosome is the better choice. It takes time,

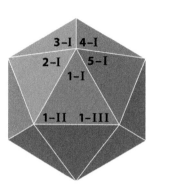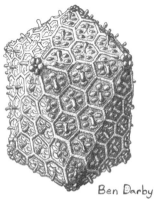

Ben Darby

Figure 40: Icosahedral variations. (Left) Yoda's capsid is a regular icosahedron composed of 60 capsid proteins – three (labeled I, II, III) in each of 20 faces (numbered 1, 2, etc., and protein pentamer at each of the 12 vertices. (Right) Lander's prolate icosahedral capsid shell contains 960 major capsid proteins arranged with pentamers at the vertices and hexamers elsewhere. Original drawing by Ben Darby, from *Life in Our Phage World* by Rohwer, F, et al. 2014. Used with permission.

energy, and amino acids to make each protein. The smaller the proteins and the fewer needed for each virion, the more virions a phage can produce from the same host resources in the same amount of time. But there is a trade-off. The smaller the capsid, the smaller the chromosome that can be transported within. A larger capsid costs more in resources, but the additional genes can provide significant competitive advantages. Phage–phage competition is fierce. Even a very small increase in production efficiency, when amplified over many generations and evolutionary time, may enable a phage to out reproduce its competitors. One way to minimize the number of protein molecules required, regardless of chromosome size, is to construct a spherical or nearly spherical capsid. A sphere has the smallest surface-to-volume ratio of any geometric solid. The smaller that ratio, the less capsid material needed to enclose the same cargo space.

The optimal capsid would also offer ease of assembly, a portal for chromosome entry and exit, and a shell that, while economically thin, is nevertheless strong enough to withstand the forces from within and without. The vast majority of phages use capsids with icosahedral symmetry. (Picture the geodesic domes designed by Buckminster Fuller.) Describing them as icosahedral does not mean that the capsids have flat faces, crisp edges, and sharp, pointed vertices. Rather it denotes the arrangement in three-dimensional space of the capsid proteins, an arrangement that creates the two-fold, three-fold, and five-fold axes of symmetry that define an icosahedron (see Figure 40 and also "Figure 4" on page 13). Being constrained by rules inherent to icosahedral symmetry, the simplest capsid is composed of 60 identical proteins. Five proteins are positioned so as to form a pentamer at each of the 12 vertices. Some phages assemble larger capsids with the same symmetry using multiples of 60 proteins. In addition, Lander and others also stretch their icosahedron into a prolate geometry. A few eccentric phages do use other, non-icosahedral architectures. These might be relics of a simpler, ancestral assembly method or perhaps more recent adaptations. In either case, today these and other architectural eccentrics constitute only a small fraction of the phages known to infect Bacteria, but are a robust portion of those with archaeal hosts.

Who Supervises?

Assembly of a mechanical apparatus or a piece of furniture in our factories is a task to be carried out intentionally by some agent (human or machine) that has been provided with a set of instructions and an input of energy. Every step introduces the possibility of error. This necessitates quality control checks post-assembly, and all too often the product fails the test. In contrast, there is no virion assembler. Phages have no need for an assembler, nor for architectural drawings or an energy source. Like Taoist sages, they practice non-doing, allowing everything to happen spontaneously. Their job is only to provide the parts needed – "smart" parts that "know" how to spontaneously assemble into a capsid containing the correct numbers of every structural protein, accurately positioned and in harmony with their neighbors. So self-managed is this process that often the parts don't even have to be inside a living cell to assemble; they can do it in a test tube. They do it seemingly effortlessly, turning out capsid after capsid with the same precise geometry and dimensions. Quality control is excellent, as we would expect for today's highly evolved phage. For Lander, Temperance, and many others, every virion is able to launch an infection.

Is this bottom-up management of a complex project unique to phage virion assembly? Looking around at the world of animals we find analogous instances where a higher order of complexity emerges from the activities of a leaderless group. We can think of each virion protein as an independent agent akin to an ant in a colony, a bird in a flock, or a fish in a school. Each of these agents individually follows a few simple rules that tell it how it is to respond to local conditions, such as whom to follow and whom to move away from, or which phero-mone trail to pursue. In the case of a capsid protein, these local rules tell it which proteins it can bind to and at what location, and in what sequence. None of these agents are aware of the bigger architectural picture, nor do they need to be. Their individual actions, governed by straightforward rules, lead to large-scale patterns and group strategies that we, as outside observers, want to attribute to some coordinating entity giving orders. These emergent properties[3] are manifestations of swarm intelligence.[4] Another example close to home is our own highly valued thinking. Neurons in our brain act as individual agents, each one releasing particular neurotransmitters from specific dendrites in response to those that it, in turn, had received from other neurons. Thinking is an emergent property of these neurons. No one neuron can think a thought, but many neurons, each one following local rules, can think about phage assembly.

During capsid assembly, the behavior of individual structural proteins is governed by the local rules implicit in their intricate three-dimen-sional form, a form that depends on the protein's amino acid sequence. Although it is necessary, having the correct amino acid sequence is not sufficient. To be functional, a newly synthesized protein must transform from a jumbled string of amino acids into a specific three-dimensional structure – the protein's "fold" (see "Protein Structure" on page 57). Only when correctly folded can a protein fulfill its enzy-matic or structural roles. Evolutionarily-related proteins that perform similar functions often have the same recognizable complex fold, even though there may be no detectable similarity between their amino acid sequences. Some proteins reliably fold themselves as they emerge newborn from the ribosome; some require help from other proteins,

---

[3] emergent property: a property of a group that does not reside in any individual member, but emerges from the combined lower level actions of the group members.
[4] swarm intelligence: the apparently intelligent collective behavior that results from the localized interactions of many individuals within a decentralized, self-organiz-ing system.

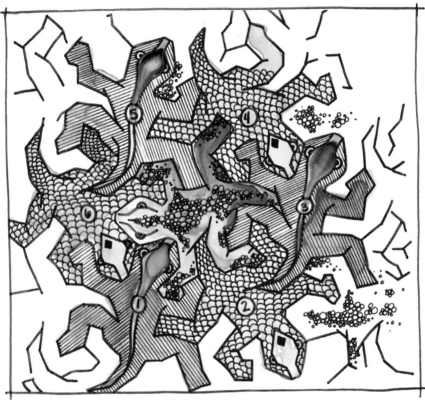

Figure 41: Thinking like Escher

the chaperones. All cells have chaperones. One type, the chaperonins, forms protected folding chambers that sequester the amino acid chain away from the bustle of the cytoplasm and provide an environment that favors correct folding. Phage proteins also may need folding assistance. For this, many phages rely on virocell chaperonins, while some, such as Lander with its large genome, also encode a few of their own.

All the proteins needed for virion assembly are synthesized concurrently late in an infection. Their rates of synthesis are regulated to produce more of those needed in greater numbers for each virion, fewer of the others (see "Management Basics" on page 90). After that, the proteins do their job without any supervision. As virion proteins accumulate in the crowded virocell cytoplasm, they are jostled about, often colliding with other cellular proteins. As their numbers swell, they are more apt to bump into one another. These collisions offer opportunities to establish more enduring associations. Each protein carries distinct binding sites exposed on its surface, sites primed and ready to adhere to the reciprocal site on other specific proteins. Picture one of Escher's tessellations, the lizards if you so choose (see Figure 41). These engaging patterns are two-dimensional arrays of identical pieces that fit together in only one particular orientation. Now picture such an array on the surface of a sphere, an Escher globe. (Phage icosahedral capsids are almost spherical.) Flesh out these lizards into three dimensions, their limbs intertwining. Convert each lizard into a capsid protein, or perhaps a trimer of capsid proteins, whose shapes and binding sites dictate where and how they connect and interlock. These linkages, governed by local rules, determine the architecture that takes shape as more proteins join the assembly.

Telltale Family Resemblances

Phages evolve rapidly. The amino acid sequence of each protein changes quickly, the nucleotide sequence encoding that protein even faster. Comparison of the same protein found in two phages will reveal overall sequence similarity only if the phages are quite closely related, in other words have recently diverged from a common ancestor. Some key amino acids may be conserved longer, such as those that form the active site of an enzyme or that are key to the folding of the protein. However, the overall protein fold is often conserved long after sequence similarity has been blurred by time. Comparing folds provides a way to look farther into the past and to see the evolutionary relationship between more distantly-related phages (see Figure 42).

This approach to phage genealogy is most accurate for the proteins that comprise the phage "self," the essential components that do not change as the phage adapts to a new host or different environmental conditions. These include the structural proteins of the capsid and the enzymes involved in virion assembly, such as the terminase.[5] Determination of the high resolution structures of some of these proteins has revealed phage family lineages whose members includes those that infect Bac-

[5] terminase: a motor protein that uses energy from ATP to translocate linear dsDNA into a preassembled capsid shell (procapsid).

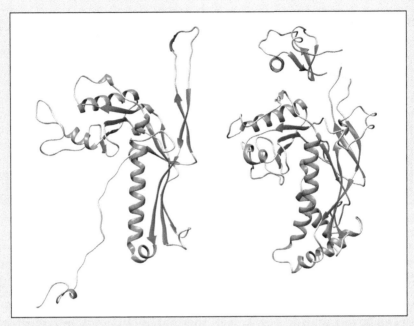

Figure 42: A conserved major capsid protein fold. The fold of Lancelot's major capsid protein (the HK97 fold) is found in numerous other phages, as well as in eukaryotic viruses. These ribbon diagrams represent the three-dimensional structure of the folded major capsid proteins of Lancelot and Lander, both of which, like most tailed phages, use the HK97 fold. The spiral regions indicate α-helical secondary structure, the flat arrows denote β-strands, and the connecting ropes are random coil loops. Courtesy of Janne Ravantti, Institute of Biotechnology and Department of Biosciences, University of Helsinki, as a service of EU Instruct ICVIR operation.

teria, Archaea, and even Eukarya (including us). A compelling interpretation of this is that these phages have a common ancestor that was present before the three domains of life diverged from one another. Thus, these family resemblances provide a time machine that we can use to probe the earliest stages in the evolution of life on Earth.

What Makes It Go?

Virion proteins are not covalently linked to one another. Instead virion assembly and durability relies on many weak interactions such as those from the attraction and repulsion of charged amino acids. Also

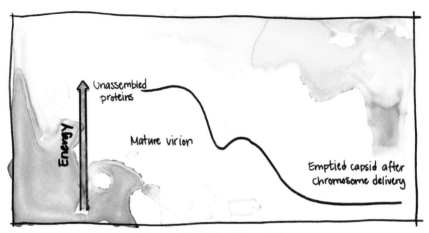

Figure 43: Going Downhill

contributing to the bonding are the hydrophobic, uncharged amino acids that associate with one another, eager to hide from water molecules. Because all of these types of protein-protein bonds are weak, they can be more readily broken by thermal fluctuations and reformed during assembly. This ability to backtrack allows small assembly mistakes to be corrected. Similarly, these individually weak bonds allow protein conformations to be modified in later assembly steps, and they facilitate the even more dramatic structural rearrangements required later for chromosome delivery into the host cell (see "Special Delivery" on page 215). Despite their individual fragility, the sum of many weak interactions provides adequate virion stability and resilience for the hazardous extracellular journey.

A general principle of phage virion assembly – the principle of prerequisites – guides the sequence of assembly. Step two can be undertaken only if step one has been satisfactorily completed, and so on down the assembly line. This efficient and fail-proof management method acts during assembly of capsids, tails, and tail fibers, and also governs the sequence of other processes such as packaging of the chromosome and the addition of external capsid decoration proteins.

What provides the energy to assemble the protein components to construct a virion? Virocell-generated ATP is available, but it is not needed. Assembly does not require an input of energy, thus fits the scientific definition of a spontaneous process. To visualize spontaneous and non-spontaneous physical processes, we often imagine an energy hill analogous to a familiar physical hill on Earth (see Figure 43). Due to gravity, a ball spontaneously rolls downhill, moving from higher to lower potential energy states. We have to input energy to push it back up. Similarly, due to the potential energy in their intra- and inter-molecular bonds, unassembled capsid proteins are higher on the energy hill than when assembled into a capsid. Assembly is spontaneous; virions do not spontaneously fall apart because disassembly means disrupting many weak bonds, thus would be going up a significant hill.

A stable virion is only partway down the hill, not at the bottom. Given the appropriate trigger, it can move on downhill to an even more stable state. Thus, a virion is more precisely described as metastable. This second transition is also a spontaneous process in that it does not require an energy input. The trigger is adsorption to a potential host cell. Adsorption triggers the virion proteins to rearrange into a lower energy, more stable conformation. This rearrangement is accompanied by delivery of the phage chromosome into the cell.

How Is It Done?

Caveat

Here I must reiterate a caveat: every generalization I offer about the phages has exceptions. Every phage "rule" is broken by a few, perhaps many. I have resorted to some generalizations so as to succinctly convey some key strategies. All generalizations are oversimplifications.

Faced with the task of assembling a virion containing at least 60 proteins—often hundreds—how does a phage proceed? There are two general approaches. Plan A is to first form subassemblies of five or six proteins and then assemble the capsid from these larger building blocks. With Plan B, each capsid protein is added individually to the growing capsid shell. In either case, a scaffolding protein usually assists. The first step seems easy. Individual proteins associate with other proteins. However, the bonds are weak and the proteins can drift apart again driven by random thermal energy fluctuations. How to encourage these clusters, once formed, to persist long enough to co-alesce into something larger and more stable? Assembly proceeds in steps. When each step is completed, the new structure is locked to prevent disassembly. The locking can result from conformational changes of the proteins that yield a lower energy, more stable configuration. Frequently steps can't be taken out of sequence because Step 2 requires the product of Step 1, and so on all the way down the assembly line.

When assembling icosahedral capsids via Plan A, the capsid proteins first assemble into clusters, or capsomers,[6] forged by connecting capsid proteins to form five- and six-membered rings (pentamers and hexamers). Joining a ring shifts each protein to a more stable configuration. Being slightly lower on the energy hill, the rings resist disassembly. They don't grow larger because all the protein-protein binding sites on the member proteins are already occupied. The juxtaposition of proteins within them creates the new binding sites that enable capsomer rings to link to one another–the next assembly step. As capsomer links to capsomer, the shell architecture that we admire takes shape. The position and orientation of these connections enable the linked capsomers to form a thin closed shell that is only one protein thick.

Two other protein types also play a part: portal proteins and scaffolding proteins. As their name implies, the portal proteins form the channel through which the chromosome enters and exits the capsid. Often twelve portal proteins preassemble into a dodecameric ring that then nucleates[7] the assembly of a new capsid. This ensures that, for the phages that use portals, each capsid will have a portal at one and only one vertex. Exceptions to this include Yoda (ϕX174) who needs no portal and Slick[8] who creates a distinctly different type of portal complex (see "Slick Assembly" on page 148). The other players here are the scaffolding proteins.[9] By binding the assembling proteins and guiding them into position, these scaffolds make the process go more

---

[6] capsomer: a subunit of a phage capsid comprising a hexameric or pentameric ring of capsid proteins. Often capsomers pre-assemble from individual capsid proteins, then self-assemble with other capsomers (and sometimes a portal) to form the capsid.

[7] nucleate: (verb) to facilitate the first step in the formation of a structure, such as a capsid, by self-assembly of its component parts.

[8] Slick: phage PRD1, a tectivirus.

[9] scaffolding protein: a protein that guides capsid assembly but is not present in the mature virion.

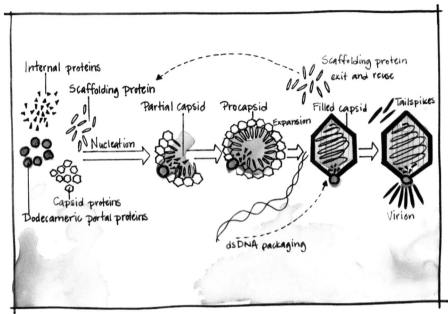

Figure 44: Chimera's assembly pathway. Late in infection, the capsid, scaffolding, and portal proteins accumulate. Twelve portal proteins preassemble into a ring that nucleates assembly of a capsid. Individual coat proteins, escorted by a scaffolding protein, assemble into a capsid shell with an interior scaffold. Packaging of the DNA drives the scaffolding proteins out through procapsid pores and expands the procapsid. The ejected scaffolding proteins are reused to guide assembly of the next capsid.

rapidly and ensure that the finished capsid has the correct dimensions. Sometimes the same capsid proteins can, given different scaffolding, assemble into two different size capsids – a property profitably exploited by Thief (P4; see "The Thief" on page 138). When the chromosome has been packaged inside the capsid, the no longer needed scaffolding proteins are shed, sometimes to be reused in assembly of another capsid.

Chimera[10] uses the one-by-one method referred to as Plan B (see Figure 44). Each of Chimera's capsids comprises 420 capsid proteins arranged as 12 pentamers and 60 hexamers. As in Plan A, scaffolding proteins – here more than 300 – guide the conformation of the growing capsid. Capsid proteins alone in a test tube do not assemble, nor do the scaffolding proteins when alone. But mix the two together, and – pres-

---

[10] Chimera: enterobacteria phage P22, a podovirus (based on its morphology) that infects Gram-negative *Salmonella*.

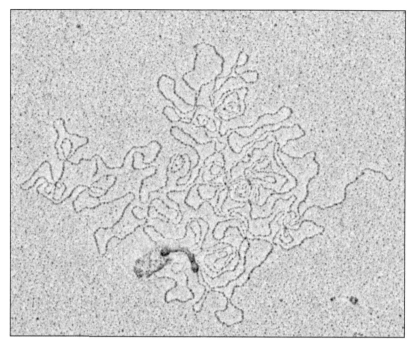

Figure 45: DNA spaghetti. The length of a phage chromosome is far greater than the dimensions of its capsid. *Listeria* phage P70, shown here after release of its DNA, packages its 67,170 bp dsDNA chromosome inside a prolate icosahedral capsid 128 nm long and 57 nm diameter, then attaches a 141 nm long tail. Courtesy of Jochen Klumpp, ETH, Zurich, Switzerland.

to!–the capsid proteins assemble to form closed icosahedral shells, each of which surrounds an inner scaffold. *In vivo*, the portal proteins preassemble first. Then individual capsid proteins, each escorted by a scaffolding protein, accrete onto the growing structure. Both Plan A and Plan B lead to the same end: a procapsid[11] of the right dimensions, ready to receive the phage chromosome.

Packaging DNA

At this point, assembly has produced a protein shell called a procapsid. Much more work lies ahead. Proteins that the phage needs to have at hand immediately on arrival, proteins such as the restriction endonuclease (RE) inhibitor used by Stubby (T7), must be loaded on board. If internal scaffolding proteins were used during assembly, they are

---

[11] procapsid: a virion assembly intermediate, specifically an assembled icosahedral capsid prior to chromosome packaging.

expelled before or while the DNA enters. The chromosome itself must be transferred into the waiting procapsid. Consider the magnitude of this task (see Figure 45). For example, Lander's 169 kbp chromosome is a dsDNA molecule 52 μm (52,000 nm) long. All that DNA needs to fit inside a prolate icosahedral capsid that is 115 nm long and 85 nm wide. That is comparable to stuffing a thin rope (~4 mm diameter) the length of a soccer field into an undersized soccer ball through a hole about twice the diameter of the rope. In the process, the chromosome is compressed ten thousand-fold compared to its volume in an aqueous solution. Little Dynamo's[12] chromosome is about a tenth the size of Lander's (19 kbp), but so is the volume of its 54 × 42 nm prolate capsid. A similar challenge is faced by almost every phage with a dsDNA chromosome. Of necessity, the DNA inside the capsid must be highly organized and densely packed, perhaps ordered like thread wound on a spool.

Chromosomes of dsDNA defy such treatment. The molecule's helical structure resists excessive bending, and if two regions of the long molecule are brought close together, the negative charges along the phosphate backbone in one region will repel the negative charges in the other. As a result, DNA resists being densely crowded. To overcome its resistance, phages have evolved ATP-fueled packaging motors, terminases, that are among the strongest biological motors known. These terminases generate forces 20–25 times greater than that of our muscle protein, myosin. They also consume ATP at a rapid rate – one ATP for every two base pairs packaged for some phages, higher in others. Equipped with this powerful pump, phages package dsDNA to a very high density, to approximately the density of liquid crystalline[13] DNA (500 mg/ml). Some phages also package a thousand or more proteins into the capsid for delivery along with the chromosome.

The amount of DNA packaged varies greatly among different phages, but so too does the capsid volume. The net result is that the DNA density inside different size capsids, with or without associated proteins,

---

[12] Dynamo: *Bacillus* phage φ29, a podovirus.
[13] liquid crystal: a physical state intermediate between a liquid and a crystal in which a substance may flow like a liquid while its molecules remain in a stable crystalline array.

is quite similar. Perhaps this density represents the maximum that a motor protein such as terminase can achieve, or perhaps it is limited by capsid strength. High DNA density increases the osmotic pressure[14] inside the capsid to forty, sixty, or even more atmospheres[15] – approaching the pressure experienced 2000 feet below the surface of the sea. These two forces – the repulsion due to the charged DNA backbone and the osmotic pressure – threaten to rupture the thin (2–4 nm) capsid wall. To resist these pressures, some phages resort to adding "glue proteins" to hold their capsids together.

Most often dsDNA phage chromosomes are packaged for transport as a single linear molecule. This may sound straightforward, but often is a rather complicated affair. Synthesis of daughter chromosomes during an infection requires a circular template. Thus, the chromosome must switch from linear in the virion to circular in the virocell and then back to linear in the progeny virions. When first made, new daughter chromosomes are linear molecules, not circular, which solves part of the problem. However they are made as concatemers: continuous strings of typically five or more genomes arranged head-to-tail. These strings must be cut into genome-length molecules for packaging. Each packaged daughter chromosome must also be equipped with a way to recircularize on arrival. The various packaging strategies devised by the phages satisfy both requirements.

Enter the Terminase

By the time packaging gets underway, procapsids and DNA concatemers are accumulating in the virocell cytoplasm. Each procapsid is equipped with one specialized vertex that serves as the portal for DNA

---

[14] Suppose you have a closed container separated into two compartments by a semi-permeable membrane. One compartment you fill with water with a low concentration of salt, the other with a more concentrated salt solution. Water, but not salt, can pass through the membrane. The difference in salt concentration between the two compartments generates a force (the osmotic pressure) that drives water from the low salt to the high salt compartment. If the strength of the walls of the high salt compartment is less than that force, those walls will rupture. The analogous situation exists when DNA and other solutes are more concentrated inside than outside a capsid. The capsid shell must be strong enough to resist this osmotic pressure.

[15] "Atmosphere" was historically defined as a unit of pressure equal to the mean atmospheric pressure at mean sea level at 15° C. This is a significant amount of force. Although we are usually oblivious of it, the weight of the atmosphere above us is 14.7 lbs/inch² or 1.03 kg/cm².

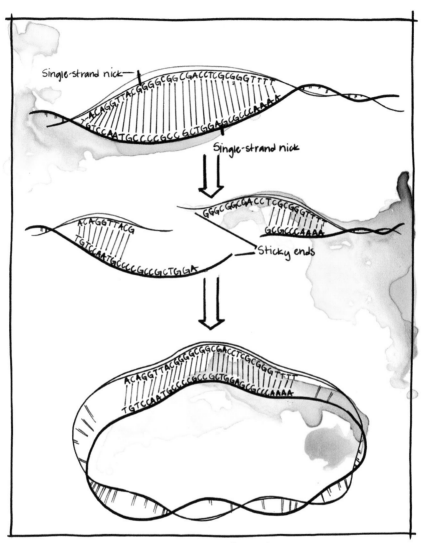

Figure 46: Sticky Ends

passage. Here the usual capsid proteins are replaced by twelve copies of the portal protein arranged in a ring. In the center is a hole large enough to allow the passage of a dsDNA molecule. This is the doorway through which the DNA enters during packaging and later exits during infection. Also on hand at this time are copies of terminase, the multi-functional enzyme that does the work of packaging. A terminase protein locates and binds to a specific short DNA sequence near the free end of a concatemer. Still holding onto that DNA, it docks to the portal of a procapsid. Once docked, it starts to pump the DNA into the procapsid. As the procapsid's cargo space becomes more crowded, the work gets harder and the rate of filling slows. When the terminase determines that the procapsid has been filled, it cuts the packaged DNA free from the rest of the concatemer. In the final packaging step, a few head completion proteins are added at the portal. Some of these form a plug, known as the gatekeeper, that stabilizes the completed capsid and prevents premature DNA ejection. Meanwhile, the terminase, still holding the concatemer by its newly cut end, moves off in search of the next procapsid to fill.

How does the terminase know when it has packaged enough DNA? Most often it uses one of two strategies. Both strategies make the cuts in such a way that the packaged linear chromosome will recircularize immediately upon arrival. Recircularization is essential not only for replication, but to protect the chromosome from host exonucleases (see "Death by Nuclease" on page 64).

Temperance,[16] for example, encodes a specific packaging sequence in its chromosome, the *cos* site, that signals to the terminase where to cut. As the terminase pumps the DNA into the procapsid, it cuts at each *cos* site. By proceeding this way, every chromosome that Temperance packages is identical. They are all the same length, and they all start and end at the same points. To prepare for recircularization later, terminase makes two nicks for each "cut," one in each strand, twelve base pairs apart (see Figure 46). This creates a single-stranded overhang of twelve nucleotides at both ends of each chromosome. The two "sticky" ends of each chromosome are complementary. After arrival in the host cytoplasm, the overhangs at the two ends base pair with each other and the nicks in each strand are sealed by a host enzyme. This generates a circular chromosome ready for replication.

The alternative strategy is a bit sloppy, but it also works. Here, too, the terminase initially recognizes a packaging sequence in the phage chromosome and makes a cut nearby. Likewise, it threads the free end into a waiting procapsid and starts pumping the DNA inside. But this time it continues until it senses that the capsid is full. It cuts the pack-

---

[16] Temperance: enterobacteria phage λ, a siphovirus that infects *E. coli*.

aged chromosome free from the concatemer at that point and heads off to repeat the process with the next procapsid. How does the terminase decide when the "head," or capsid, is full? Perhaps it senses the seismic effects of packing DNA to such high density inside the capsid. Packaging irreversibly expands the capsid, typically doubling its volume. The capsid wall thins and acquires a more angular profile. As packaging nears completion, distortion of the portal might give the docked terminase its cue to cut and run.

This "headful" packaging method is imprecise. Instead of cutting at a specific site to complete each chromosome, the terminase guesstimates where to cut. Thus, every chromosome starts and ends at a slightly different point. To ensure that every virion contains a complete genome, terminase packages a bit more than a single genome length, in the range of 2% to 10% more. This extra length serves another purpose, as well. No matter where the cuts are made, the gene(s) at the chromosome's leading end are duplicated at the trailing end. This redundancy enables the chromosome to circularize on arrival. Homologous recombination between the duplicate regions converts the extra-long linear chromosome into a circular chromosome containing precisely one genome. The extra DNA is discarded.

How long does it take to fill a procapsid? That depends on chromosome length and the terminase's packaging speed. Lander (T4) holds the current world record for DNA packaging speed: 2000 bp/second. If it could package its entire 169 kbp genome at that rate, it would be finished in less than 90 seconds, but it can't. The rate slows markedly as packaging proceeds due to the increasing resistance from the crowded DNA. Averaged over the entire packaging process, Lander's rate falls to ~700 bp/second, still fast enough to package its large genome in four minutes. In general, phages with larger DNA genomes (like Lander) have faster motors, those with smaller genomes have slower motors, with the net result that almost all phages complete packaging in two to five minutes.

Terminases package linear chromosomes of dsDNA through a packaging portal located at a specialized vertex. Phages with different chromosome types – linear ssDNA, ssRNA, or dsRNA; circular dsDNA or ssDNA – must rely on other methods. A few use a different type of DNA translocase to package through a temporary opening located in a face of the icosahedral capsid rather than at a vertex (see "The Yoda Way" on page 142). Others eliminate the need for an energy-demanding packaging step by not preassembling a procapsid. Instead their capsid assembles around the chromosome (see "Live and Let Live... and Exploit" on page 149).

Terminase packaging is reliable. However, given the gazillion phage virions packaged, errors do occur. Occasionally the terminase grabs a piece of host DNA by mistake and packages it into a capsid. This creates a mature virion that heads out into the world carrying bacterial, not phage, genes. Like their siblings, when these virions encounter a potential host cell they adsorb and deliver their cargo. But instead of launching an infection, these virions shuttle genes between closely related bacteria. Although infrequent, these mistakes have been evolutionarily important (see PIC).

Lancelot's Habergeon[17]

Temperance's cousin Lancelot,[18] knighted in honor of its superb armor, adds its own special twists to an otherwise typical assembly and packaging protocol. Its medium-sized (55 nm diameter) capsid assembles from 415 capsid proteins and twelve portal proteins. Although a capsid of this size requires scaffolding, Lancelot proves that this does not require dedicated scaffolding proteins. Its capsid proteins could be said to be not only self-assembling, but also self-scaffolding. When assembled into a procapsid, the terminal 102 amino acids at one end of each capsid protein form a long arm that protrudes into the interior. Interactions between neighboring arms provide the scaffolding that guides capsid assembly. The procapsid shell when first assembled is completely filled by these arms along with about 50 proteases.[19] As soon as shell assembly is finished, the proteases go to work chopping the scaffolding arms, and themselves, into fragments that exit through the pores in the procapsid shell.

---

[17] habergeon: a short, sleeveless shirt of chain mail.
[18] Lancelot: phage HK97, a siphovirus.
[19] protease: an enzyme that cleaves the peptide bonds that link one amino acid to another in a polypeptide or protein. Also known as a peptidase or proteinase.

While this type of scaf-
folding is rare, targeted
cutting of every capsid
protein after initial as-
sembly is a common
practice that confers several benefits. First, it allows newly made capsid
proteins to be optimized
for assembly, then sub-
sequently be modified to
better function in the mature capsid. Second, the
cleavage of a protein by a
protease is an irrevocable
act that makes assembly
irreversible. There is no
going back, no disassem-

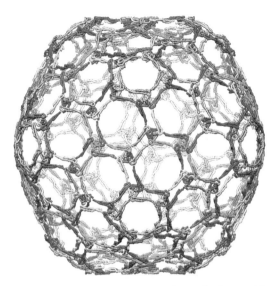

Figure 47: Lancelot's armor. Lancelot's capsid
proteins intertwine to form a strong, resilient catenane. Courtesy Gabriel Lander, The
Scripps Research Institute.

bly allowed. Lastly, if only cleaved capsid proteins are able to engage
in the next assembly action, steps can not occur out of sequence. Many
phages use this tactic to keep the assembly line moving smoothly forward from start to finish.

Once the proteases have done their job and exited the procapsid, chromosome packaging can proceed. Packaging expands Lancelot's capsid
from 55 nm to 65 nm diameter, which almost doubles its volume and
thins the shell to a mere 1.8 nm (18 Å). This creates yet another challenge, one faced by many phages. How can such a thin-walled structure made of proteins held together by only weak bonds withstand the
high internal pressure generated by the packaged DNA? Some phages
rely on the increased capsid strength that results from conformational
changes of the capsid proteins during expansion. Others attach many
copies of a "decoration protein"[20] to the exterior to glue together adjacent
capsomers. This glue is essential for Temperance (λ). Without it, the high
internal pressure (more than 60 atmospheres) would rupture its capsid.

---

[20] decoration protein: a structural protein added to a capsid after assembly that may
reinforce the capsid, alter the overall electrical charge of the capsid, facilitate binding to various surfaces, or serve other purposes.

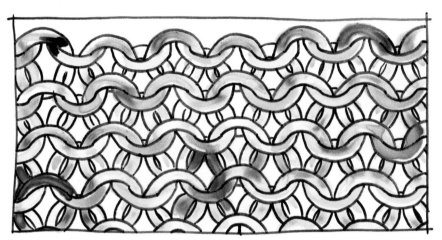

Figure 48: The catenane pattern used for the traditional European 4-in-1 weave.

Lancelot has a similar capsid with a similar problem, but has a different solution. As its capsid expands during packaging, all 400-plus capsid proteins simultaneously shift position and intertwine their polypeptide arms to "stitch" together neighboring capsomers. In addition, each capsid protein in every capsomer ring has a particular arginine[21] facing its neighbor protein on one side and a lysine[22] facing its neighbor on the other side. As these proteins shift position, the arginine of each one is brought close to the lysine of its neighbor in a microenvironment that catalyzes spontaneous formation of a covalent bond between them. These bonds covalently link each protein to its two neighbors like a group of people in a circle holding hands. To now disrupt Lancelot's capsid would require breaking these covalently-linked rings. Moreover, due to the intertwining, these rings are topologically interlocked like the links in a chain (see Figure 47). This simultaneous interplay over the entire capsid generates one continuously interlocked protein catenane or, in other words, chain mail (see Figure 48). Humans figured out this interlocking trick, too, and used catenanes to fabricate flexible armor in both Europe and Japan. However, our armor is a clunky, cumbersome, and costly added layer. In elegant contrast, Lancelot's sleek chain mail is not a second skin, but an intrinsic property of its only skin.

---

[21] arginine: one of the amino acids that typically carries a positive charge *in vivo*.
[22] lysine: one of the amino acids that typically carries a positive charge *in vivo*.

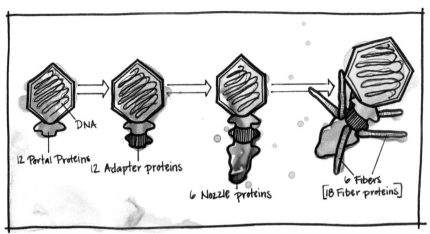

Figure 49: Growing a stubby tail. Stubby's tail grows at the virion portal by the sequential assembly of 48 protein molecules.

Three Phage Tales

Capsid assembly, filling, and expansion completes virion construction for some phages, but the most successful phage families go one step farther: they add a tail. Although bacterial phage tails come in three distinctive forms, they all have two properties in common. They all carry fibers or spikes that recognize a potential host on contact, and they all form a tubular channel through which the dsDNA chromosome will exit the capsid. Even the simplest of tails adds some complexity to virion assembly and increases the cost of virion production. The benefits become apparent as we watch the phage secure its next host (see "The Quest" on page 185) and deliver its chromosome (see "Speculative Tails" on page 222).

Stubby (T7) appears to have the simplest tail, merely a short, nozzle-like structure attached at the capsid portal (see Figure 49). This short tail looks inadequate for the task of infecting its *E. coli* host, a task that requires the tail to span the 24 nm periplasmic moat that separates the cell membrane and the outer membrane of this Gram-negative bacterium. Appearances can be deceiving. Hidden inside Stubby's capsid are all the component parts needed to assemble a tail extension during DNA delivery (see "Delivery with a Stubby Tail" on page 227). Stubby's visible external tail is a modeststructure as phage tails go. It is assembled from only four different proteins, 48 protein molecules in all, and its assembly could hardly be more straightforward. Three protein rings and some fibers are all that is needed. One ring is the dodecameric portal that is located at the unique vertex of the capsid. This portal is also called the connector because, as you guessed, it connects the tail to the filled capsid. It also provides binding sites for twelve copies of the adapter protein that join the structure one by one and form the second ring. Lastly, six copies of the nozzle protein attach to the adapter ring to complete the tail tube. Meanwhile, stiff, 32 nm long fibers with

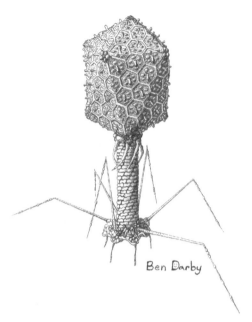

Ben Darby

Figure 50: A drawing of Lander's sophisticated virion. Credit: Ben Darby. Previously published in *Life in Our Phage World* by Rohwer, F, M Youle, H Maughan, N Hisakawa. 2014. Wholon. Used with permission.

a central kink preassemble independently. Each is a coiled-coil composed of three copies of the fiber protein twisted together over their entire length like the strands in a rope. Attachment of a fiber to each of the six nozzle proteins completes the virion. All in all, Stubby's tail grows itself, the pieces seemingly falling into place one by one, all in the correct sequence.

The flamboyant contractile tails of the larger Myoviruses, such as Lander (T4), are far more complex (see Figure 50). They announce that this is a phage with a wealth of genes and advanced assembly skills. Lander dedicates 25 kbp of its 169 kbp genome to tail production – an amount equivalent to many complete phage genomes. Of its 40 or so different structural proteins, more than half are found in its tail, a total of ~465 protein molecules per tail. Unlike Stubby's assembly-by-accretion, Lander divides the work among three independent, but concurrent, assembly lines. One subassembles the capsids, one focuses on tail

Figure 51: Lander's Assembly Lines

construction, while another puts together the long tail fibers (LTFs). A preassembled tail attaches to each capsid, and afterwards the six LTFs join to complete the virion.

The first step on Lander's tail assembly line is to put together the baseplate that forms the distal end of the tail, the end farthest from the capsid (see Figure 51). This is an exceptionally complicated structure composed of ~150 protein molecules of at least 16 different types. Sixteen may seem excessive given that many phages construct their entire virion from only a handful. However, this baseplate is pivotal when a virion initiates an infection (see "Delivery by Lander" on page 226).

After the virion docks at the host surface, the baseplate actively assists the chromosome to exit and paves the way for its entry into the host cell. To construct a new baseplate, six preassembled wedges join together around a central hub. Soon this structure is joined by three copies of the tape measure protein that each attach by one of their ends. After a few more proteins have been added, the baseplate is ready to grow the tail tube. This long tube will provide a 40 Å (4 nm) wide channel running the length of the tail through which the dsDNA chromosome and internal proteins will migrate into a host cell.

A Journey Down Lander's Tail Tube

This video (http://bit.ly/29LCPTS), created by Steven McQuinn, opens with a drawing of an assembled Lander virion, then zooms in on the baseplate at the terminus of the tail tube. The red form partially visible within is the rigid spike or needle that extends the tail tube past the baseplate. This hollow structure offers a central channel that can accommodate DNA passage, and is also sufficiently rigid to puncture the CM. The remainder of the video flies down the terminus of the tail tube and thence through the needle – the route Lander's chromosome follows to enter the cell.

How does a phage build a long, hollow tube with a constant diameter both inside and out? By stacking rings of tail tube proteins on top of the baseplate, the proteins forming circlet upon circlet around the tape measure proteins. Stacking halts when the stack reaches the free end of the tape measure. For a 100 nm long tube, this will be when there are 144 proteins in the stack. The same essential process is then repeated to assemble the outer tail sheath, also 100 nm long, from 144 sheath proteins stacked around the tail tube. In both cylinders, the proteins are stacked in a consistent helical array – a pattern widely used in the viral world to economically assemble rod-shaped structures from small, identical subunits.

When first assembled, the tail sheath is extended to the full length of the tail tube. Like a stretched spring, this sheath is in a metastable state

Figure 52: Lander virions coming and going. Assembled Lander virions have accumulated inside this *E. coli* host. Additional virions are adsorbed to the surface. Those with empty capsids have ejected their chromosome. Courtesy of John Wertz, Yale University.

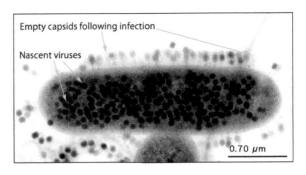

Empty capsids following infection

Nascent viruses

0.70 μm

awaiting the moment the virion adsorbs to a host cell. The tape measure proteins remain inside the tail tube, waiting, until they, too, can assist with chromosome delivery. The completed tail then connects to a prepared capsid. In the process, six stiff bristles – the whiskers – are added that project out from the "neck" that joins the "head" and the tail.

Meanwhile, completed LTFs are coming off a separate assembly line. Each LTF is composed of two rigid sticks, each of which assembled from three copies of one or two specific proteins. The sticks are joined at a 20° angle by a "knee" formed by yet another protein. Total length of an LTF? An impressive 144 nm (.144 μm), making it the longest component of Lander's virion. Six of them must be attached to each waiting capsid-plus-tail as the final step in assembly. Imagine the challenge of maneuvering these long, skinny rods in the crowded cytoplasmic gel of the virocell, a cell that is only ~.5 μm wide and ~2 μm long. Moreover, one end (and not the other) must be brought to a binding site on the baseplate. For this, Lander cannot rely solely on random collisions as it does for many other assembly steps. To increase the probability of success, Lander uses its whiskers. A whisker temporarily holds onto a completed LTF near the knee. Doing this brings the correct end of the LTF into position at the baseplate. (This grasping of each LTF by a whisker persists after assembly and assists during the extracellular search for a new host. See "Landing on the OM with a Long Tail" on page 198). One by one, six tail fibers attach to the baseplate. The finished virion then joins its siblings awaiting the moment of their release into the world outside.

Quality control? Flawless. Essentially every Lander virion can launch an infection (see Figure 52).

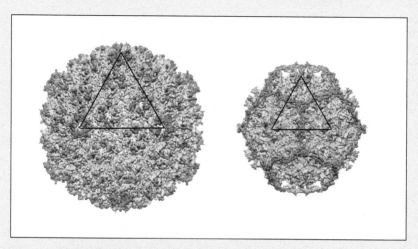

Figure 53: Will build to suit tenant. Surface view of cryo-EM procapsid reconstructions of Thief's helper phage, P2 (Left) and Thief (Right). The triangles drawn on the images denote one face of the icosahedral procapsids. Image resolution is 9.5 Å for Thief and 9.9 Å for the helper. Courtesy of Terje Dokland, University of Alabama at Birmingham.

The overall process of tail acquisition is similar for both families of long-tailed phages: the myoviruses such as Lander and the siphoviruses including Lancelot (HK97). The flexible siphovirus tail begins with the assembly of an initiator structure analogous to Lander's baseplate. Here, too, three (or perhaps six) tape measure proteins attach next. They provide the scaffolding for the growth of the tube to the correct length. Addition of a puncturing spike at the end completes the naked tail tube. The tape measure proteins remain inside and are essential for DNA delivery during infection.

The Thief

Phages exploit other phages. One ploy used is to steal your structural proteins from a "helper" phage. Such thievery is not simple since the helper has its system in place for assembling and filling capsids with its own DNA. A successful thief needs to be on the scene at the right time and then see to it that capsids are filled exclusively, if possible, with its own DNA. A phage thief can't steal from just any phage. Such a coup rests upon an intimate relationship.

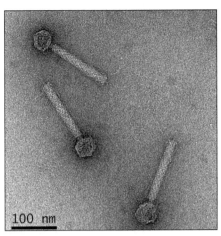 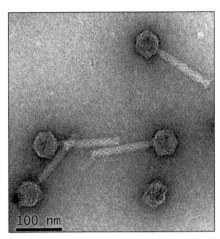

Figure 54: Thief and helper. TEMs of Thief (Left) and its helper phage (Right) highlighting the difference in their capsid dimensions. Courtesy of Terje Dokland, University of Alabama at Birmingham.

Thief[23] has evolved a very sophisticated scam. Since it relies on its helper phage[24] to provide the genes needed for virion assembly, chromosome packaging, and virocell lysis, Thief does not need as many genes. Compared to its exploited helper, it can make do with a chromosome that is only one third as long. It takes advantage of this difference by stealing the helper's capsid proteins and assembling them into Thief-sized capsids, capsids that are too small for the helper's chromosome (see Figure 53). Perhaps this should not surprise us because many diverse phages build capsids of greatly differing dimensions from similar capsid proteins. Thief simply takes this one step farther. It demonstrates that even the same capsid protein can be coaxed to reliably assemble into capsids of two distinct sizes (see Figure 54). When unmolested, the helper phage assembles 415 copies of its capsid protein into a 62 nm diameter shell. When Thief intervenes, 235 copies assemble into Thief-sized, 45 nm capsids. What determines capsid dimensions are the scaffolding proteins. The helper relies on its internal scaffolding proteins to guide assembly of its larger capsid. Thief synthesizes its own scaffolding protein that works in conjunction with the helper's scaffolding to form the small capsids. Thief's scaffold forms an external net that tethers the assembling capsomers, constraining them

[23] Thief: phage P4, a siphovirus that infects *E. coli*.
[24] Thief's helper is phage P2, a myovirus that infects *E. coli*.

Build Your Own Lander Virion

for phage phanatics of all ages

Ben Darby

Everything you need to assemble your own scale model of a Lander virion enlarged five million times. Assembled, it stands a stately 1.8 m (71 inches) tall from capsid apex to the tips of its extended tail fibers. A dramatic statement for your office or living room!

Comes with a complete set of structural proteins ready to assemble. No instructions or tools required. To assemble the interlocking transparent components, simply match up the color-coded, protein-to-protein binding sites using the adhesive tabs provided. Each protein comes with reusable adhesive tabs that allow you to disassemble to correct errors. Made from 100% recycled bacterial cytoplasm. Hours of entertainment for young and old alike.

Order Yours TODAY!

DIFFICULTY RATING: Advanced.

ASSEMBLY TIME: A few minutes for a phage. Humans typically take considerably longer.

ASSEMBLY TIP: Let the parts be your guide. Your instructions, including the sequence of assembly, are implicit in the parts themselves. Listen to their advice. You do not know better than the phage.

NOVICE KIT: 4381 proteins; capsid, tail, tail fiber, and scaffolding components boxed separately.

MASTER ASSEMBLER KIT: 4381 proteins all jumbled together in one box for added realism.

Optional DNA packaging module: includes 169 kbp of DNA helix to scale (250 m, 820 feet), 1100 internal proteins to be packaged with the DNA, and battery-powered terminase robot programmed to dock at the portal and spool the entire DNA rope inside the capsid. Amaze your friends with this billion-year-old packaging feat!

A phantastic project for families, clubs, and zealous individuals!

LANDER KIT
Parts List

To order your kit:
www.thinkinglikeaphage.com

Optional holiday add-on package: includes 160 miniature lights that insert into the capsid to replace the standard Hoc decoration proteins, a Santa-style hat for the capsid apex, precut tinsel to wrap the collar and whiskers, and six hand-knit knee-high stockings of recycled organic Icelandic wool to slip over the LTFs.

For beginners, we suggest our cheerful Yoda mobile. Kit includes enough interlocking proteins to assemble six icosahedral capsids in rainbow colors. Capsid diameter 15 cm (6 inch). Comes complete with hanger and strings. Hang one above a crib or in a child's room to inspire the next generation of phage researchers. Safe for all ages. All parts are made from non-toxic materials that, if accidently ingested, will dissolve and promote a healthy gut microbiome.

Qty	Description (part name or number)
Capsid components	
Note: Included are the scaffolding proteins and others that you need temporarily to facilitate assembly.	
960	major capsid protein (23)
12	portal (20)
55	vertex (24)
840	stabilizing decoration (Soc)
160	decoration (Hoc)
115	internal scaffold (22)
3	internal scaffold (21)
370	internal scaffold (IPIII)
360	internal scaffold (IPI)
360	internal scaffold (IPII)
40	internal portal (Alt)
240	internal, temporary (68)
341	internal, temporary (67)
Tail	
18	whiskers (Wac)
6	wedge (53)
12	wedge (6)
6	wedge-vertex (7)
12	wedge (8)
18	wedge-vertex (9)
18	wedge-pin (10)
18	wedge-pin (11)
6	wedge (25)
6	likely wedge (Frd)
3	hub (5)
3	hub (27)
3	likely hub (Td)
18	short tail fibers (12)
6	baseplate (48)
6	baseplate (54)
144	tail sheath (18)
144	tail tube (19)
6	tail tube (29)
6	tail terminator (15)
6	tail tube terminator (3)
Long tail fibers	
18	proximal section (34)
6	hinge region (35)
18	inner distal section (36)
18	outer distal section (37)

to assemble into a smaller capsid. These mini-capsids are less stable than their larger counterparts, so Thief adds a stabilizing decoration protein, as well.

Even if Thief snatched every capsid protein and assembled them into small capsids, the helper might still try to package its chromosome inside. None of the resultant virions would be infectious because they would not contain a complete helper chromosome. Nevertheless, every capsid that is filled with any helper DNA is one less capsid available for Thief. To prevent this, Thief takes advantage of terminase specificity. A terminase will package any DNA that contains the appropriate short packaging sequence – a tactic that invites exploitation. All a robber needs to do is mimic that sequence in its own DNA. Actually, Thief does even better. Its chromosome has a packaging sequence that is even more inviting to the helper's terminase, so inviting that the terminase preferentially packages Thief DNA. Of course, successful thievery requires the helper to be providing the structural proteins when Thief needs them. As you imagine, Thief has this handled, too, as well as what to do when there is no helper in sight. This intricate story of phage-phage interactions will be explored further in PIC.

The Yoda Way

Although tailed phages with dsDNA chromosomes packaged by powerful terminases are the majority among the known phages, there are numerous other ways to assemble a virion. Consider Yoda (ϕX174), a very different sort of phage that is considered by some to be the most elegant and the most sophisticated of them all. Its circular chromosome of ssDNA is much smaller – 5,386 nt compared to dsDNA chromosomes that are often in the range of 30,000 to 200,000 bp. It has, in one respect, the simplest of architectures (see Figure 55). Its icosahedral capsid is also small, only 27 nm in diameter, and it uses the minimum number of major capsid proteins – 60 proteins arranged as 12 pentamers. But Yoda does not stop here. It adds a pentamer of another structural protein at each of the 12 vertices to form the projecting spikes vital for host recognition and adsorption. That adds 60 more structural proteins to each virion. Note that all of the vertices as they assemble are identical; there is no specialized portal vertex for chromosome packaging.

Immediately upon arrival, Yoda recruits host proteins to convert its ssDNA chromosome into the circle of dsDNA needed to serve as a template for both transcription and chromosome replication. Early in the infection, this dsDNA molecule is repeatedly replicated and the needed structural proteins are synthesized. Sixty individual capsid proteins assemble into capsomers; these capsomers, already associated with 60 spike proteins (5 per spike), assemble into procapsids guided by external scaffolding proteins. At this stage, there is no capsomer-

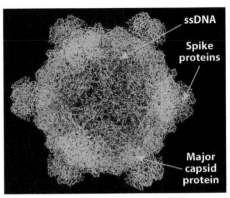

Figure 55: Yoda's virion. Structure of Yoda's virion determined to 3.4 Å resolution by X-ray crystallography. Courtesy of Fdardel (http://bit.ly/2elV8pv). Generated from PDB entry 2BPA. Original publication: McKenna, R, D Xia, P Willingmann, LL Hag, S Krishnaswamy, MG Rossmann, NH Olson, TS Baker, NL Incardona. 1992. Atomic structure of single-stranded DNA bacteriophage ΦX174 and its functional implications. Nature 355:137-143.

to-capsomer binding. Instead the capsomers are held in place by two groups of scaffolding proteins, one internal and the other external – a novel strategy that may account for Yoda's very rapid rate of assembly. Later in infection, the needed enzymes, together with individual circular dsDNA templates, dock to a procapsid and begin to synthesize the linear ssDNA that will be packaged. One of the replication enzymes uses energy supplied by ATP to feed the available end of the newly-made DNA strand into the procapsid. This continues until a specific site on the chromosome is reached, at which point a cut is made to free the completed chromosome from the replication complex. The chromosome ends are joined to form a circular ssDNA chromosome inside the capsid. Sixty DNA-binding proteins accompany the DNA into the procapsid, in the process driving out the internal scaffolding proteins. The external scaffolding proteins are released and the capsid collapses around the ssDNA, its diameter decreasing by almost 2 nm.

A closer look confirms that Yoda's packaging process is markedly different from the established paradigm. The tailed phages all use a powerful packaging terminase to pump their DNA into a procapsid through a specialized portal vertex. The result is an expanded and densely packed virion. Yoda uses a DNA replication enzyme to thread a newly-synthesized strand of DNA into the procapsid through one of the pores in the 20 icosahedral faces. The procapsid then collapses around the small ssDNA chromosome. The 60 DNA-binding proteins associated with the packaged DNA are rich in basic amino acids, thus have a net positive charge. By binding to both the negatively charged DNA and the inner surface of the capsid they tether about 10% of the DNA close to the protein shell. Yoda's method for delivering its chromosome into a new host is also, not surprisingly, extraordinary (see "Yoda's Grand Entry" on page 229).

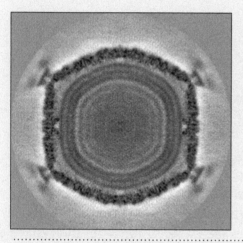

Figure 56: Fusion's virion. A cryo-EM cross section through Fusion's complex virion reveals spikes at the vertices and, moving inward, the outer protein shell followed by two layers of protein-rich lipid membrane enclosing the spooled DNA. Credit: Image created by Pasi A. Laurinmäki from the PM2 electron microscopy density map (Electron Microscopy Data Bank EMD-1082). http://bit.ly/21hQUNz

A Lipid Supplement

All the virions considered so far are constructed from proteins and nucleic acids exclusively, but some eccentric phages include lipids in their capsids. Both Slick (PRD1) and Fusion[25] enclose their chromosome within a membrane sac inside the usual protein shell (see Figure 56). For Slick in particular, the associated membrane and protein layers endow the capsid with exceptional strength and resilience—a property that we also use in numerous composite materials. Shy (φ6) wraps its entire viri-

---

[25] Fusion: phage PM2, a corticovirus that infects *Pseudoalteromonas*.

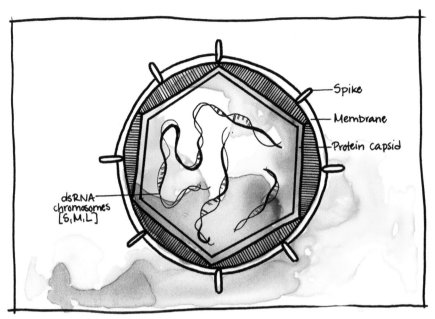

Figure 57: Shy's Virion

on within a membrane in the final step of assembly. All of the lipids needed to form these membranes are stolen selectively from the virocell. Having an internal membrane complicates virion construction, but having a membrane – inside or out – opens the door to some ingenious methods of chromosome delivery. All of these phages take full advantage of this opportunity during infection (see "Lipid Opportunities: Conduit Construction" on page 233, "Lipid Opportunities: Membrane Fusion" on page 230, and "Lipid Opportunities: Metamorphosis" on page 235). Lipid membranes, both internal and external, are more common among the phages infecting Archaea. So far, little is known about either their assembly or their role in adsorption or escape.

Shy's Packaging Feat

Shy belongs to a small phage family that boldly defies the customs of the majority. Not only are they the only family to encode their genome in dsRNA, but they divide their dozen or so genes among three separate chromosomes (S, small; M, medium; and L, large; see Figure

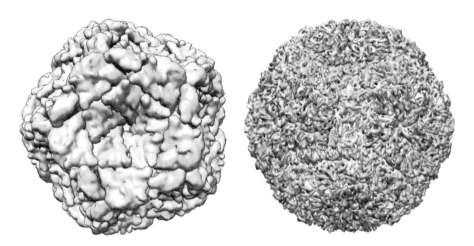

Figure 58: Shy's packaging challenge. Cryo-EM reconstructions of (Left) Shy's procapsid (4.4 Å resolution) and (Right) Shy's capsid after packaging and addition of the outer capsid, but before enclosure within the external membrane (7.5 Å resolution). Source: Protein Data Bank (Left, ID 4BTG; Right, ID 4BTQ). Primary publication: Nemecek, D, E Boura, W Wu, N Cheng, P Plevka, J Qiao, L Mindich, JB Heymann, JH Hurley, AC Steven. 2013. Subunit folds and maturation pathway of a dsRNA virus capsid. Structure 21:1374-1383.

57). This genomic choice creates two major challenges. Delivering your genome as dsRNA invites attack by the cell's endonucleases. Shy's solution is to take its capsid into the host for protection (see "Shy's Delivery" on page 232) and to allow only single-stranded transcripts to venture outside the capsid throughout the infection. It packages its chromosomes as ssRNA and then adds the complementary strand to each one within the confines of its capsid shell. Second, Shy must ensure that one, and only one, copy of each chromosome is packaged in each virion. At first glance, this may seem tricky and error-prone, but it is readily handled by a "thinking" phage. Shy employs the principle of prerequisites to ensure that each chromosome is packaged in sequence.

After arrival but still inside the capsid, one strand of each of Shy's dsRNA chromosomes serves as the template for synthesis of ssRNA transcripts that exit to the cytoplasm. Initially these transcripts are translated into proteins, including the capsid proteins. The capsid proteins assemble into procapsids, as usual, but Shy's procapsids are distinctly not round (see Figure 58). Picture a partially deflated soccer ball that has been punched in at many places so that, rather than pointing outward, each vertex is sunken inward. The space inside is empty and small, but also expandable. Each of the twelve vertices contains a channel large enough for ssRNA to pass. Each vertex is ringed by six packaging motor proteins on its exterior face. On the interior face, slightly offset so as to not block passage, sits an RNA-dependent RNA polymerase (RdRP) that, when given a strand of ssRNA, can synthesize the complementary strand.

As the infection proceeds, S, M, and L ssRNA transcripts that exited into the surrounding cytoplasm are available for packaging. Each transcript contains an essential packaging sequence, ~200 nucleotides long, located near one end – different sequences for S, M, and L transcripts. These differences enable the ssRNAs to form specific stem loops and other secondary structures that are unique to each type of transcript. As a result, the three transcript types are each distinctive in both shape and sequence. Shy utilizes these differences to determine which chromosome should be packaged at each step. The surface of an empty, collapsed procapsid offers binding sites that recognize and bind the packaging region of S transcripts – only S transcripts. An S transcript attaches and is imported into the procapsid by the packaging motor at the vertex nearby. This expands the procapsid slightly, enough to make the S binding sites disappear and the M sites form. An M transcript is then packaged the same way, further expanding the procapsid. Lastly the L strand is brought inside. This final expansion activates the RdRPs embedded within the capsid. These enzymes synthesize a complementary strand for each of the packaged transcripts, thus completing the dsRNA genome – one dsRNA copy of each chromosome reliably packaged in each virion.

Shy's virion now has a full complement of chromosomes, but the capsid is not finished. Again deviating from the phage majority, it adds a second, outer protein shell. It then wraps its entire virion in a lipid membrane that is derived from the virocell membrane and that is also rich in embedded phage proteins. That membrane later serves to unlock the door for entry into the soon-to-be host cell (see "Shy's Delivery" on page 232).

Slick Assembly

Instead of an external envelope, Slick (PRD1) carries its membrane inside its virion where it closely follows the contours of the icosahedral protein shell. Having an internal membrane raises many questions. Does Slick assemble its protein shell first and then add the lipid lining? Or does the protein shell assemble brick by brick on the outside of a membrane sac? And how does Slick's dsDNA chromosome pass through the membrane on its way both in and out of the capsid? The DNA action occurs at the unique virion vertex. Whereas each Slick virion has 11 vertices equipped to recognize and adsorb to a host cell, it is the twelfth vertex that organizes the assembly and later serves as the site of DNA entry and exit. As with the tailed phages, capsid assembly is nucleated by the future portal. This is another example of the principle of prerequisites, in this case used to ensure that there is only one "unique" vertex per virion. However Slick's portal is more complex than that of the tailed phages, being built from four different phage proteins instead of one. Assembly begins when two of them, organized as six heterodimers, arrive at the virocell membrane. These are the inner portal proteins that will form a 4–5 nm wide DNA conduit through the phage membrane. A patch of the CM, likely with embedded phage membrane proteins, invaginates and pinches off to form a vesicle or sac. As procapsid assembly proceeds, those inner portal proteins are joined by dodecamers of each of the two outer portal proteins, one dodecamer on top of the other, together forming a pore through the protein shell. The outer dodecamer is also the DNA packaging motor. Combined, all four proteins form a channel that penetrates both the protein shell and the membrane.

Packaging of Slick's chromosome also differs from the typical tailed phage method. Instead of departing from the portal after packaging as does terminase, this packaging motor is a structural component of the mature virion. Also, DNA packaging does not cause upheaval and expansion of the procapsid shell, but does expand the membrane vesicle and push it closer against the capsid shell – ready to play a part during chromosome delivery (See "Lipid Opportunities: Conduit Construction" on page 233).

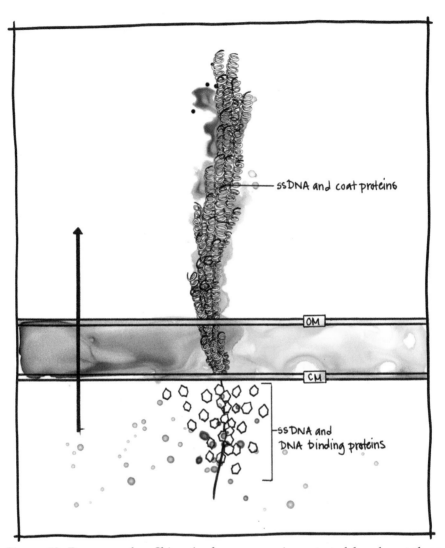

Figure 59: Dragon scales. Skinny's chromosome is protected by almost three thousand overlapping coat proteins in the finished virion.

Live and Let Live...and Exploit

Not all phages repay their host with lysis. The majority of those infecting Archaea eschew lysis, especially those in extreme environments, but little is known about their alternative exit routes. One group of phages[26] that infect Bacteria has been observed to continually siphon off resources and release virions without killing the virocell. Skinny

---

[26] the family *Inoviridae*.

(Ff) embodies this lifestyle. Its small, ssDNA chromosome (6,407 nt) contains only ten genes, five of which encode virion structural proteins. New chromosomes are synthesized continuously and are quickly enshrouded by ssDNA binding proteins. Skinny also synthesizes its structural proteins continuously and parks them temporarily in the membrane, poised for service. Unlike the case for the vast majority of phages, completed virions do not accumulate inside the virocell awaiting the moment of their joint escape. Virions are not assembled in advance, but instead as each chromosome individually slithers through the virocell membrane. On the way out, the ssDNA-binding proteins are replaced by 2700 copies of the waiting coat protein (see Figure 59). These replacement proteins overlap like fish scales to protect the chromosome. In addition, a different pair of specialized proteins are attached at each end. The completed protein coat displays helical geometry, a pattern reminiscent of the long tails of Lancelot and Lander, but these filamentous virions are far longer and skinnier – 850 nm long and a mere 4.3-6.3 nm in diameter. For comparison, this is approximately eight times longer than Lander's tail but only one-fourth as wide.

The existence of Skinny's family proves that their tactics can work, but their paucity in the phage world attests to its limitations. Although they do not kill their virocell, they do divert cell resources for their own reproduction. This is not an insignificant drain, as Skinny can produce a thousand progeny per virocell generation. The budding of so many virions through the cell surface also interferes with the formation of other surface-penetrating structures, such as those used for motility. Moreover, this mode of virion construction imposes severe constraints on the phage. One obvious constraint is limited chromosome length. The far greater cargo capacity of icosahedral capsids allows room for both a longer chromosome and internal proteins. Skinny's virion length and ssDNA chromosome make intercellular travel riskier, the former by increasing the risk of virion damage and the latter by increasing susceptibility to UV inactivation (see "*Hic Sunt Dracones*" on page 188). Although it is possible for a phage to carry out a replicative life cycle with ten genes or even fewer, the larger and more diverse phage families routinely have chromosomes ten to forty times longer. More genes offer more opportunities for self-defense, host manipulation, a broader host range, and coping with varying environmental conditions.

Figure 60: An extremophile's capsid protein. Ribbon diagram of Biped's major capsid protein that forms both the spindle and the two tails. Unlike most capsid proteins, this one is a bundle of α-helices. The spheres represent chloride atoms complexed with the protein when crystallized for X-ray diffraction. Courtesy of Protein Data Bank (3FAJ). To be published: Goulet, A, G Vestergaard, U 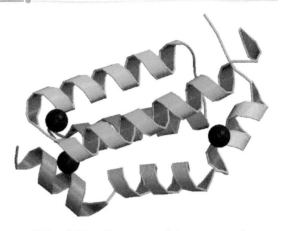 Scheele, V Campanacci, RA Garrett, C Cambillau. Structure of the structural protein P131 of the archaeal virus Acidianus Two-tailed virus (ATV).

Extracellular Assembly

The extremophilic[27] Crenarchaeota[28] inhabiting hot springs around the globe are host to a zoo of phages. Their virion architectures include the familiar icosahedral capsid, both with and without a tail, but here these forms are the minority. There are also rods and filaments that may superficially appear familiar but that have their own unique characteristics. And then there are imaginative forms that are unique to this group, forms such as spindles with or without tails, bottles, and droplets. Like their hosts, these phages thrive at extreme high temperatures.

Biped[29] actually needs its preferred temperature of 85° C to complete virion assembly in a timely manner. Its virion is a spindle, about 243 nm long and with a 119 nm girth at the middle. What sets Biped apart from its spindle-shaped relatives is its two long tails that give it an average tailspan of 744 nm. These "tails" are not independently assembled structures tacked onto a preassembled icosahedral capsid, such as we saw earlier for Stubby and Lander. Rather these tails grow as extensions of the spindle-shaped capsid. Each tail is a thin-walled tube ending in 15-25 nm long tail fibers. Both the spindle and the tail

---

[27] extremophile: an organism that thrives in environments with extremely high or low temperature, pressure, salinity, or pH (acidity or alkalinity).

[28] Crenarchaeota: a major phylum within the domain of the Archaea that includes many hyperthermophiles (organisms that thrive at temperatures of 80° C and above).

[29] Biped: archaeal phage *Acidianus* Two-tailed Virus, a bicaudavirus.

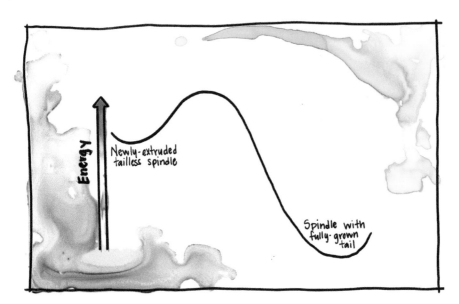

Figure 61: Over the hill. Biped's tail growth is a spontaneous reaction. However, while en route to the final lower energy state, the virion must first surmount an energy hill. The height of the hill relative to the starting state represents the activation energy required for the process to proceed.

assemble from the same capsid protein (see Figure 60). To assemble a virion with these properties and capabilities requires a major capsid protein distinctly different from the ones used by most phages. Compare the fold of Biped's protein shown here – a bundle of four long α-helices – with the conserved fold of Lancelot's and Lander's proteins (see "Figure 42" on page 119).

Assembly of a virion with a tailspan of 744 nm poses challenges. Recall that special attention was required for Lander to maneuver and attach its LTFs during assembly inside a viscous, crowded virocell, and those LTFs were only 144 nm long. Biped has its own solution, one that is unique among all known phages so far. Its virions extrude from the virocell as spindles with pointed ends, lacking anything you'd call a tail. The tails grow after exiting. In the process, the total volume of the virion – spindle plus tails – decreases about 50%. We don't know exactly how Biped manages this, but we have some clues. The extruded virions are self-sufficient in that they carry every molecule needed for this transformation, including the tail components. Timely tail growth does require an energy source in the form of thermal energy. At Biped's usual environmental temperature in the range of 85° C., tails grow rapidly and are completed in less than an hour. If experimentally held at 4° C, the spindles remain tailless for months. Thus, the virions are in a metastable conformation when they emerge from the virocell. They subsequently transition down the energy hill to a more stable, lower-energy, tailed configuration (see Figure 61). However, this spontaneous transformation requires high temperatures. The virions have to first go uphill before they can slide down to the more stable state. That uphill climb is the activation energy. Biped harnesses the thermal energy available in its normal environment to power this climb. It may also be using ATP as the energy source for part of the conformational change. One structural protein is predicted to have the needed capability, and it is possible that Biped packages some of the virocell's ATP pool inside its spacious virions. Possible, but untested.

A Two-Way Street

Theft and exploitation goes both ways. Phages steal intellectual property from their hosts, and hosts have co-opted phage technologies. Capsid technology, for example, is useful for a cell, as well as for a traveling phage. Major capsid proteins have been adapted and used to build cellular nanocompartments to house bacterial enzymes. Why do this? Often a metabolic pathway involves several enzymes that act in sequence, one handing off its product to the next for further processing. Line them up inside a compartment with pores that allow the substrate to enter and the product to leave, and the cell has a more efficient assembly line. Moreover, some metabolic intermediates are toxic to the cell. Sequestering them inside a nanocompartment protects the cell from harm. Similarly, sensitive enzymes or metabolites inside the compartment are protected from damage wreaked by other cellular components.

Why stop with only in-house use? Capsids are superb vehicles for cargo delivery to targeted cells. To an enterprising prokaryote, a virion offers a pre-fabricated vehicle for horizontal gene transfer[30] to a related cell. Diverse prokaryotes – ~6% of them overall – have taken advantage of

---

[30] horizontal gene transfer: the movement of one or more genes from the chromosome of one prokaryote to the chromosome of another cell of the same or a different prokaryote species (or to a eukaryote).

this by securing control of virion production and then filling the capsid with cellular DNA. Although no longer an infectious phage, these gene transfer agents[31] still rely on phage-derived components and assembly mechanisms. They also retain their original host specificity, thus ferry genes between closely-related cells. Because, like their phage progenitors, they depend on cell lysis for release, their production is suicide. Thus, deployment is tightly regulated and restricted to only a small percentage of the population and only in times of major stress.

Numerous prokaryotes have discarded the phage capsid and co-opted the tail instead. Proteins that make up a secretion system used by bacterial pathogens to inject nasty proteins into target cells show clear signs of phage ancestry. Other Bacteria brazenly use modified, but clearly recognizable, phage tails (tailocins[32]) to attack specific target cells. Some tailocins kill by simply puncturing the cell membrane, while others are loaded with toxic cargo that is delivered into the cell. For more on the evolutionary and ecological ramifications of these stories, see PIC.

Without an overarching blueprint or supervision, newly minted virion parts assembled themselves into exquisite forms. Architectural patterns emerged as each component linked to its neighbors. A quiescent phage chromosome was packaged into each procapsid along with a few proteins also destined for delivery. Completed virions came off the assembly line, each one prepared to defy environmental assaults but also responsive to contact with a potential host. As production continued, virions accumulated, all waiting to leave home together and set out in search of a new host of their own. Time now to escape!

---

[31] gene transfer agent: a virus-like particle that is produced and released by a prokaryote, and that contains cellular DNA.
[32] tailocin: a co-opted phage tail, encoded by and produced by a prokaryote, that upon release targets and kills sensitive cells.

Further Reading

Abrescia, NG, DH Bamford, JM Grimes, DI Stuart. 2012. Structure unifies the viral universe. Annu Rev Biochem 81:795-822.

Aksyuk, AA, MG Rossmann. 2011. Bacteriophage ssembly. Viruses 3:172-203.

Aoyama, A, RK Hamatake, M Hayashi. 1983. In vitro synthesis of bacteriophage phi X174 by purified components. Proc Natl Acad Sci USA 80:4195-4199.

Casjens, SR, EB Gilcrease. 2009. Determining DNA packaging strategy by analysis of the termini of the chromosomes in tailed-bacteriophage virions. in *Bacteriophages*: Springer. p. 91-111.

Chen, D-H, ML Baker, CF Hryc, F DiMaio, J Jakana, W Wu, M Dougherty, C Haase-Pettingell, MF Schmid, W Jiang. 2011. Structural basis for scaffolding-mediated assembly and maturation of a dsDNA virus. Proc Natl Acad Sci USA 108:1355-1360.

Christie, GE, T Dokland. 2012. Pirates of the Caudovirales. Virology 434:210-221.

Duda, RL. 1998. Protein chainmail: Catenated protein in viral capsids. Cell 94:55-60.

Fokine, A, MG Rossmann. 2014. Molecular architecture of tailed double-stranded DNA phages. Bacteriophage 4:e28281.

Hafenstein, S, BA Fane. 2002. φX174 genome-capsid interactions influence the biophysical properties of the virion: Evidence for a scaffolding-like function for the genome during the final stages of morphogenesis. J Virol 76:5350-5356.

Häring, M, G Vestergaard, R Rachel, L Chen, RA Garrett, D Prangishvili. 2005. Virology: Independent virus development outside a host. Nature 436:1101-1102.

Heinhorst, S, GC Cannon. 2008. A new, leaner and meaner bacterial organelle. Nat Struct Mol Biol 15:897-898.

Huiskonen, JT, F de Haas, D Bubeck, DH Bamford, SD Fuller, SJ Butcher. 2006. Structure of the bacteriophage φ6 nucleocapsid suggests a mechanism for sequential RNA packaging. Structure 14:1039-1048.

Jardine, PJ, DL Anderson. 2006. DNA packaging in double-stranded DNA phages. in *The Bacteriophages*: Oxford University Press. p. 49-65. http://bit.ly/2jkMt5o

Lang, AS, O Zhaxybayeva, JT Beatty. 2012. Gene transfer agents: Phage-like elements of genetic exchange. Nat Rev Microbiol 10:472-482.

Leiman, P, S Kanamaru, V Mesyanzhinov, F Arisaka, M Rossmann. 2003. Structure and morphogenesis of bacteriophage T4. Cell Mol Life Sci 60:2356-2370.

Marvin, D, M Symmons, S Straus. 2014. Structure and assembly of filamentous bacteriophages. Prog Biophys Mol Biol 114:80-122.

Oksanen, HM, MM Poranen, DH Bamford. 2010. Bacteriophages: Lipid containing. eLS. http://bit.ly/2j5L4yx

Rao, VB, M Feiss. 2015. Mechanisms of DNA packaging by large double-stranded DNA viruses. Annu Rev Virol 2:351-378.

Chapter 5.

Escape!

In which

the assembled progeny virions depart in concert, each setting out in search of a host of its own. The time of their departure is regulated by the infecting phage, not by the waiting children or the virocell. Most phages prepare an exit route by digesting a region of the peptidoglycan cell wall, a few block peptidoglycan synthesis. Without strong wall support at all points, the virocell explodes and spews out the virions along with its guts. Some phages create more elegant egress portals, while the progeny of yet others slip out without destroying the virocell that had nurtured them.

Old Coli never die, they just phage away.
Source: Cold Spring Harbor Phage Course 1952.
http://bit.ly/1TydNGP

The timing of death, like the ending of a story, gives a
changed meaning to what preceded it.
Mary Catherine Bateson

There is a time for living and a time for dying, a time for
planting and a time for reaping, a time for motion and a time
for stillness, a time for working and a time for rest.
Carol Deppe, *Tao Te Ching*

You have to escape to survive, as you must survive to escape.
Adam Rapp

I realized if I didn't just go, I'd never go. Going was the key.
It didn't matter where I was headed just as long as I was
headed somewhere.
Jayden Hunter

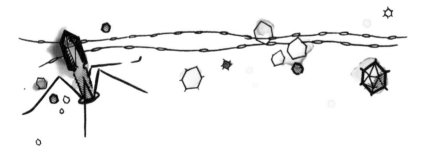

Completed virions, each a perfect copy, tumble off the assembly line and join their accumulating brethren. Meanwhile, the virocell goes on with routine business, looking normal from the outside. Perhaps it is an *E. coli* with Temperance or Cowboy on board. The virocell continues to swim briskly propelled by its flagella, fueled by continuing uptake and utilization of food. Suddenly its motion halts, the first outer sign of the internal drama. Within a minute, the virocell explodes and spews out its guts, including the waiting virions. Virocell lysis is the finale to a successful lytic infection. It plays out an estimated 10^{24} times each second on Earth. It is occurring in your gut as you read these words.

A Different Story

Infection with intent to lyse is not the only script followed by infecting phages. Many choose to collaborate with the bacterium, to cohabitate long term as a virocell (see "Coalition" on page 241). Two genomes can be better than one. Such virocells grow and divide, generation after generation, until some disaster prompts the phage to desert the soon-to-sink ship, lysing it on the way out. Only a few phages, such as Skinny (Ff), are never guilty of lysis. Skinny finds it more

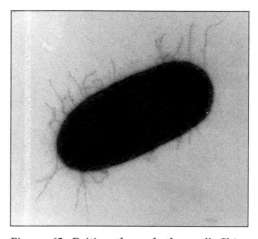

Figure 62: Exiting through the wall. Skinny's (Ff's) long, thin virions assemble as they pass through the membranes of infected *E. coli* virocells. This EM image caught extruding virions at various stages in assembly. The completed virions are up to 1 µm long, a length that exceeds the width of the virocell itself. Credit: Jasna Rakonjac. Reproduced from her doctoral thesis, with permission.

profitable to continually bleed off virocell resources for intense virion production, while continuously releasing progeny (see Figure 62 and "Scoring a Hole-in-One" on page 172).

When to Lyse?

It is simple to kill your virocell, but it is not simple to kill your virocell expertly. When done most skillfully, phage production continues at top speed right up to the moment of lysis and the progeny are sent on their way immediately afterwards – no point in hanging around in the virocell carcass when a world of potential hosts awaits you. The time of lysis, i.e., the duration of the latent period, is generally the only variable that a phage lineage can adjust to compensate for changing conditions. Lysis time is not dictated by exhaustion of host resources. When, for example, lysis by Temperance (λ) is experimentally blocked, virions can continue to accumulate to ten times or more the usual burst size. Although the lysis timer is under phage control, it is not set to go off when a particular number of impatient virions have assembled, nor is it set to maximize the number of virions produced per infected virocell. The optimal lysis time is the one that maximizes progeny production over the long term. To that end, the expert phage perpetually fine-tunes the time of lysis in response to the fluctuating availability of hosts and other resources.

The phage needs to weigh the relative advantages of two basic strategies. It can delay lysis to extend the latent period and thereby gain a larger burst size. The drawback here is that meanwhile the already assembled progeny virions sit idle. Alternatively it can lyse earlier. The burst size will be smaller, but the virions will set out sooner in search of a host. If hosts are abundant and the virions score quickly, there will soon be many second generation virocells producing virions instead of only one. For example, imagine a phage that lyses its host after 50 minutes to release 100 progeny. A competing, almost identical phage delays lysis until three hours after infection, when it releases ten times as many progeny, i.e., one thousand. Which phage wins the numbers game? It depends. A deciding factor is how many of the progeny find new hosts, and how quickly they do that. Suppose that hosts are abundant and half of the progeny successfully infect a host within ten minutes. In this case, the 50 minute phage wins. At three hours, when the procrastinator releases its 1,000 progeny, the total virion count for the 50 minute phage and its generations of descendants will be 2.5×10^5. This victory reflects the difference between a linear rate of virion pro-

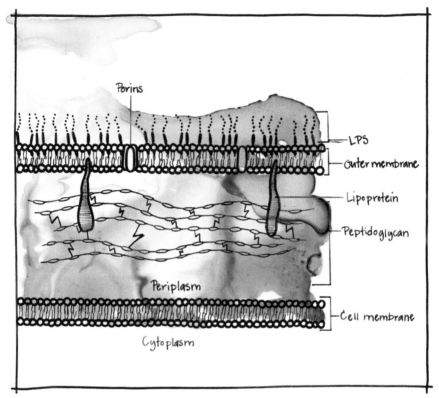

Figure 63: The nvelope of a Gram-negative bacterial cell.

duction by one virocell versus exponential phage replication. If hosts are scarce and only 2% of the progeny quickly find a home, the early lysis strategy of the 50 minute phage would yield only 400, making the procrastinator the winner.

How does a phage decide what is the optimal lysis time? As a population, they are continually testing variants with different lysis times that arise by mutation. Those variants whose time is best suited to current conditions produce the most virions in the short term. Over the following generations, their progeny come to account for a larger percentage of their local population–but only until conditions shift and favor those with a different latent period. This experimentation is forever ongoing. Bacterial populations swell or shrink, and likewise the number of competing phages will vary. Through it all, the expert phage adapts quickly.

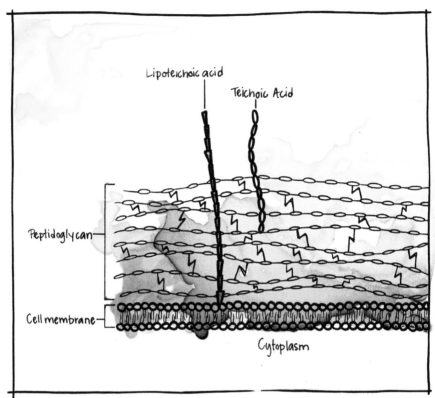

Figure 64: The envelope of a Gram-positive bacterial cell.

Gram-Negative, Gram-Positive, or None of the Above

Most Bacteria can be classified in one of two large groups based on the response of their cell envelope to a classical staining procedure, the Gram stain. On this basis, they are referred to as Gram-negative or Gram-positive. The cell membrane (CM) of those that stain Gram-negative is reinforced by a thin wall of peptidoglycan surrounded by a second membrane, the outer membrane (OM; see Figure 63). The peptidoglycan-containing zone between the two membranes, known as the periplasm, is a metabolically active cellular compartment that contains nucleases and other enzymes. Bacteria that stain Gram-positive typically have a CM surrounded by a thick cell wall of peptidoglycan (see Figure 64). In both Gram-negative and Gram-positive Bacteria, the elastic layer of peptidoglycan provides the structural support needed to counter the turgor pressure.[1] The OM of Gram-negative bacteria also contributes some resilience.

The OM is a distinctive membrane that differs in both composition and function from the CM. The outer leaflet of the OM lipid bilayer contains lipopolysaccharides (LPS) that effectively block entry of hydrophobic compounds into the periplasm. As the name suggests, LPS consists of both lipids and sugars. The lipid portion is anchored in the membrane and it, in turn, anchors a long chain of sugars – sometimes more than one hundred – that extends outward from the OM. Porin[2] proteins embedded in the membrane form channels spanning both leaflets that allow many small molecules, including needed nutrients, to pass freely into the periplasm.

Archaea share the basic prokaryote structure with the Bacteria, but differ in the details of their cell envelope. A close look reveals significant biochemical differences in the composition of their CM and also their peptidoglycan. Only some of the Archaea have a peptidoglycan layer, but all have an outer protective layer called the surface layer (S-layer[3]). S-layers are also found in some Bacteria, but aren't visible in most TEMs because the layer does not survive the typical sample preparation procedure. These layers are crystalline arrays of identical proteins, only one molecule thick (see Figure 65).

The details of these surface components matter a great deal to the phages. During infection, they provide the means for host recognition and virion attachment. They also present barriers that must be penetrated for chromosome delivery. When it is time for the progeny to escape, the structure of the cell envelope constrains the possible mechanisms for virocell lysis or virion extrusion. For those few phages that incorporate lipids into their

---

[1] turgor pressure: in prokaryotes, the pressure that continually pushes the CM against the cell wall. This force results from the inward flow of water into the cell driven by osmotic pressure, and is countered by the resistance of the intact cell wall.

[2] porin: proteins embedded in the outer membrane of Gram-negative Bacteria that span the membrane and form pores that allow the free passage by diffusion of selected small molecules.

[3] S-layer: a paracrystalline, monomolecular layer of identical proteins that comprises the outermost envelope layer of some prokaryotic cells.

virions, the composition of the virocell membrane determines what lipids are available for incorporation into phage membranes during virion assembly.

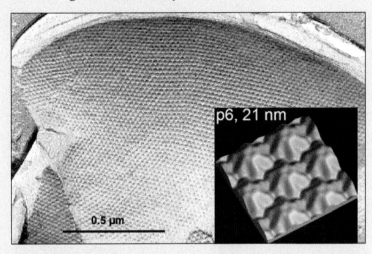

Figure 65: Archaeal protection. Scanning EM of the freeze-etched surface of the archaeon *Pyrodictium abyssi* showing the crystalline protein lattice of the protective S-layer. Inset: magnified lattice section revealing the subunits spaced 21 nm center-to-center with hexagonal (p6) symmetry. Courtesy of Reinhard Rachel, University of Regensburg Centre for EM, Regensburg, Germany.

Breaching the Great Wall

What does it take to lyse a virocell? The phage does not have to power the explosion, it only has to light the fuse by disrupting the cell envelope. The cell's turgor pressure does the rest. Turgor pressure is a direct result of cellular composition and organization. Most of a prokaryotic cell is occupied by the fat nucleoid with its structured chromosome. Surrounding that is a rich gel of macromolecules, mostly proteins, accompanied by the far more numerous small metabolites and ions that are more concentrated inside than outside the cell. The boundary between inside and outside is the cell membrane (CM; also known as the cytoplasmic membrane or the plasma membrane), a semi-permeable membrane made of lipids and proteins. Traditionally these membranes are pictured as phospholipid bilayers, but they contain more protein than lipid. The CM grants free passage to water, but not much else. Diverse molecules and ions are selectively transported into or out of the cytoplasm by a suite of channels, active transporters, and secretory systems. The avid active uptake of many molecules from the surroundings concentrates them inside the cell, which gives the cell a higher osmotic concentration than that of the surrounding milieu. This difference drives the migration of water into the cell. This, in turn, causes the cell to swell and push outward against the CM: the turgor pressure. The force of this push is usually in the range of 0.5 to 3 atmospheres (up to 20 atmospheres in some cases), a puny amount compared to the 60 atmospheres of internal pressure pushing outward against some virion capsids. Nevertheless, it is enough to rupture a naked CM.

Tomographic Visualization of Cell Rupture

A movie (http://go.nature.com/2fMfpjY) available online (provided you have a subscription or institutional access to the journal *Nature*) captures an instant in the death of an infected Gram-negative cyanobacterium (*Synechococcus*). First shown is a series of 54 Å slices from the tomographic series, which is then followed by a volume rendering that portrays the three-dimensional structure of the same region. Lastly, you can view an annotated color version with cell components and phage capsids/virions labeled. Empty capsids of infecting phages remain on the cell surface; inside the cell are assembled procapsids (both before and after DNA packaging) and mature virions awaiting release. Source: Dai, W, C Fu, D Raytcheva, J Flanagan, HA Khant, X Liu, RH Rochat, C Haase-Pettingell, J Piret, SJ Ludtke. 2013. Visualizing virus assembly intermediates inside marine cyanobacteria. Nature 502:707-710.

Gram-positive and Gram-negative Bacteria, as well as Archaea, all live enclosed within a seamless sac – the cell wall.[4] As the turgor pressure pushes outward against the CM, the elastic cell wall pushes back, and the forces balance. If a phage weakens the cell wall sufficiently, the wall

---

[4] Some Bacteria, such as *Mycoplasma*, that live within the hospitable osmotic environment inside other cells, lack a cell wall and rely on a strengthened CM.

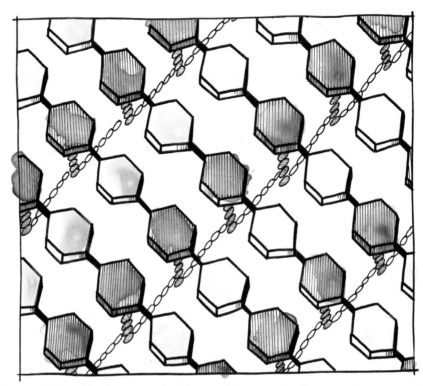

Figure 66: Peptidoglycan mesh. A layer of the peptidoglycan mesh composed of parallel glycan chains (sugars drawn as hexagons) cross-linked at regular intervals by short peptides (chains of amino acids).

is unable to counter the outward force and the virocell bursts open. This is the tactic used by most phages to release their progeny virions.

Cell walls are fashioned of multiple thin layers of a biopolymer, peptidoglycan[5] (see Figure 66). As the name indicates, it is a polymer containing chains of amino acids (peptides) and simple sugars (the glycans[6]). Both of these components differ for the most part from the amino acids and sugars commonly present in the cytoplasm. The particular sugars and peptides used vary from one bacterial group to another, and they all differ from those used by the Archaea. The first step

[5] peptidoglycan: a molecular mesh composed of chains of specific sugars cross-linked by short chains of amino acids (peptides) that forms the prokaryote cell wall. The composition of peptidoglycan in Bacteria (murein) differs from that in the Archaea (pseudomurein).

[6] glycan: a chain composed of multiple sugar molecules that may be an independent molecule or may be attached to another molecule such as a protein or nucleic acid.

in peptidoglycan construction builds linear glycan chains of typically 20-30 sugars. Each pair of sugars is covalently linked to a short peptide chain. Picture regularly spaced peptide "arms" along the entire glycan chain. The final step in construction crosslinks some of the peptide arms on adjacent glycan chains. The result is a flexible, slightly elastic mesh with many regularly-spaced pores (typically 4 nm diameter) that are large enough to allow free passage of nutrients and waste products. In Gram-negatives, the sac is thin – at most, 10 nm thick – with only one or a few layers of peptidoglycan. In the oft-studied model bacterium, E. coli, only half of the possible peptide crosslinks are made, which increases the size of some pores and allows occasional passage of larger molecules. This sac is "stapled" to the surrounding OM, an arrangement that also contributes some structural strength to the cell envelope.

By contrast, Gram-positive bacteria that have no OM build a more robust cell wall from multiple layers of peptidoglycan joined by more numerous cross-links. Here, too, the peptidoglycan mesh provides structural strength, but it makes up less than half of the material in these cell walls. The rest of the wall is composed of a complex polymer of sugars, an alcohol, and phosphate that is known as teichoic acid.[7] Some of the teichoic acids are linked to the CM, while others are attached to the peptidoglycan and extend outward beyond that layer. Total wall thickness in these Bacteria ranges from 30 to 50 nm. These walls can account for one quarter or more of the weight of the cell – a major investment of energy and resources. In neither group do these Great Walls prevent phage infection. Moreover, they are the Achilles' heel exploited by phages to bring about cell lysis.

Defects in the fabric of the walls do develop occasionally. When a break occurs in a crosslink, for example, adjacent pores connect and create a larger hole. A moderate number of such breaks poses no danger to the underlying membrane, but too many spells cell death. If such a hole in Gram-negative E. coli grows to 20 nm diameter, the turgor pressure will force a bit of the CM to irreversibly bulge through the gap (picture an aneurysm). Lysis inevitably follows. A comparable hole in the thicker Gram-positive cell wall has the same result. Some breaks

---

[7] teichoic acid: a complex polymer of sugars, an alcohol, and phosphate found in the cell wall of Gram-positive Bacteria.

are a necessary part of bacterial life. Although the peptidoglycan sacs are elastic, they require continual remodeling to accommodate cell growth. Reshaping and augmentation are most extensive when the mother cell divides. Material must be inserted at the new poles of the daughter cells without jeopardizing wall integrity, but insertion requires breaking bonds. For this breaking, Bacteria are equipped with multiple autolysins,[8] each of which cleaves a specific bond in the peptidoglycan. This potentially destructive enzyme activity is closely regulated to coordinate strand breakage with the insertion of new material.

How to Lyse Your Virocell Using Only One Gene

This brings us to the question of how a phage can effectively compromise the virocell wall to free its virions at the right time. Some phages that infect Gram-negative bacteria take advantage of the peptidoglycan remodeling required for normal cell growth. Growth makes the virocell vulnerable to the interruption of peptidoglycan synthesis. Without production of new peptidoglycan, the cell wall weakens in key regions. To exploit this vulnerability, some phages inhibit one step or another in peptidoglycan synthesis. Inhibition has no obvious immediate effect. Despite some irregularities in the mesh, the virocell maintains its normal shape and shows no apparent disturbance. It continues to grow. Compromised regions in the mesh enlarge until one reaches the critical size at which it can no longer resist the unrelenting turgor pressure. The integrity of the OM is also compromised by the disruption of peptidoglycan synthesis. Without adequate reinforcement, the cell membrane ruptures, the cell is dead, and the virions are on their way. Phages are not the only ones who use this trick. Other organisms, including us, use the same maneuver to kill Bacteria. Familiar β-lactam antibiotics such as the penicillins and cephalosporins, which we originally isolated from fungal and bacterial producers, respectively, inhibit the last step in peptidoglycan synthesis: peptide crosslinking.

This tactic is particularly attractive to coliphages with extremely small genomes, such as Minimalist (Qβ) and Yoda (φX174), because it takes only one gene to encode an inhibitor of one enzyme catalyz-

---

[8] autolysin: in general, an enzyme that digests an organism's own cells. In prokaryotes, a murein or pseudomurein hydrolase that cleaves specific bonds in peptidoglycan.

ing a step in the peptidoglycan synthesis pathway. But how do these phages modulate their time of lysis to optimize their replication for current conditions?

For Minimalist, the time of lysis is determined by two factors: when peptidoglycan synthesis is halted and how soon afterwards lysis occurs. How quickly synthesis is blocked depends on the rate of accumulation and the effectiveness of the inhibitor. Having only three genes total, Minimalist does not dedicate one to specifically encode its inhibitor protein. Instead, it assigns this task to its multi-tasking maturation protein (see "The Ingenious Minimalist" on page 92), and this protein handles the job efficiently. Each available maturation protein inhibits one copy of the enzyme that catalyzes the first step in peptidoglycan synthesis. How many of these inhibitors does Minimalist need? A typical *E. coli* host contains about 400 copies of that enzyme, so theoretically complete cessation of peptidoglycan synthesis would require Minimalist to make about 400 copies of the maturation protein. As described earlier, Minimalist needs many fewer copies than this of its replicase and many more of its capsid protein. Because its ssRNA chromosome serves also as the mRNA for all of its proteins, without some wizardry, all three–capsid protein, replicase, and maturation protein–would be made in equal quantities. The wizard is the mRNA itself that folds so as to govern ribosome access to each gene. As soon as synthesis of a new chromosome commences, a ribosome attaches and starts to work synthesizing a copy of the maturation protein. Usually before a second ribosome can access that gene, the ssRNA chromosome has assumed the folded structure that severely restricts ribosome access to that gene forever after. As a result, usually only one maturation protein is made for each phage chromosome and remains associated with that chromosome throughout virion assembly. However, occasionally a second ribosome does slip in before folding is complete and a second maturation protein is synthesized. It is these extra copies that are available to serve as peptidoglycan synthesis inhibitors. The time it takes for 400 of them to accumulate determines the time of lysis.

On the other hand, if the chromosome doesn't fold quite so quickly or if its folding is slightly less effective in blocking ribosome access to that site, then more copies of the maturation protein would be made

from each new chromosome and lysis would occur sooner. Conversely, more effective shielding of that gene could delay lysis and increase
burst size. Such changes require only a single base substitution to alter
the base pairing, and consequently the folding, in that region. Such
mutations occur at a low rate all the time. In a natural environment,
the Minimalist population includes some individuals that lyse slightly
sooner, some that lyse later. If host availability is high, those individuals with an earlier lysis time will increase in relative abundance. As
conditions fluctuate, so will the proportion of phages with various lysis times.

Lysis does not occur immediately upon cessation of all peptidoglycan
synthesis. The lag period between inhibition and lysis depends on the
rate of cell growth and this, in turn, varies with environmental conditions. The more rapidly the cells are growing, the sooner the short-
age of new peptidoglycan will become critical. When well-fed in a 37°
C lab environment, *E. coli* divides every 20 to 30 minutes. Here the
lag between inhibition and lysis is about 20 minutes. Under starvation conditions when the virocells are not growing – a frequent occurrence in many environments – inhibition of peptidoglycan synthesis
has little effect.

From one perspective, this imprecise lysis timer looks like a flaw in
Minimalist's otherwise expert replication strategy. Actually, this "imprecision" is an asset as it provides automatic, real-time adjustment
in response to current conditions. It extends the latent period when
limited resources have slowed growth and virion production. Conversely, when times are good and growth is rapid, assembled virions
accumulate quickly and lysis comes sooner. Moreover, when lack of
resources slows cell growth, host abundance also likely declines. By
delaying lysis in response to slow growth, Minimalist continues to accumulate virions inside the virocell while hosts are likely to be scarce.
When conditions improve, the increase in growth rate would prompt
virion release at a favorable time, a time when potential hosts are rapidly multiplying.

Scoring a Hole-in-One

Minimalist demonstrates that a single-gene lysis mechanism not only gets the job done, but expertly adjusts the time of lysis. The time is readily adapted genetically by mutation and is modulated during an infection in response to the probable availability of healthy hosts. Nevertheless, the vast majority of phages that infect Bacteria don't settle for a one-gene method. Instead they have a lysis gene cassette composed of at least two genes, sometimes six or even more. These phages all use the same logic: digest a large enough hole in the normal peptidoglycan sac and the cell will explode due to its turgor pressure. Enzymes that cleave peptidoglycan (lysins[9]) are widely distributed. Bacteria use their own versions, the autolysins, to revamp their peptidoglycan for repairs or cell growth. Phages use their lysins at both ends of a lytic infection: lysins at the beginning to clear a path in through the cell wall to access the CM for chromosome delivery (see "Barriers" on page 221) and endolysins[10] at the end to degrade the cell wall from the inside out for virion escape. The biochemically complex peptidoglycan mesh offers many potential targets for lysin attack. Phage endolysins show a corresponding complexity with various phage endolysins targeting at least four different peptidoglycan-specific linkages.

Expert lysis requires a phage to do more than simply synthesize an endolysin. For one thing, the phage must unleash its endolysins at the right time, precisely when it is ready to lyse and discard its virocell. Until then, it doesn't want to jeopardize the virocell metabolism that is supporting its own replication. As new virions accumulate, phages such as Temperance (λ), Lander (T4), and Stubby (T7) stockpile active endolysins in the cytoplasm. There the enzymes wait, barred from the peptidoglycan by the CM. The assault on the cell wall awaits the accumulation of another protein, the holin. As their name suggests, holins make holes in the CM, holes that allow the waiting endolysins to exit the cytoplasm and quickly attack the peptidoglycan. The holins are also the timers that determine the time of virocell lysis. Every bacteriophage with a dsDNA chromosome uses a holin-endolysin one-two punch to lyse its virocell.

---

[9] lysin: any molecule that causes cell lysis, such as phage lysins that cleave bonds in peptidoglycan.
[10] endolysin: a phage enzyme released from inside the virocell that cleaves bonds in peptidoglycan, thereby lysing the cell.

Holins comprise one of the most diverse protein families known. Their amino acid sequences often show no correspondence to one another, but they are all small proteins that embed in the CM with a characteristic topology. Each protein weaves back and forth across the membrane to bury from one to three regions (transmembrane domains[11]) within the hydrophobic, lipid-rich CM. The N-terminal end of the holin projects from the CM to the outside, while the C-terminal end extends into the cytoplasm. When only a few holin proteins are embedded in the CM, the membrane remains impermeable to most small ions. The virocell maintains membrane traffic as usual by actively transporting some specific ions into the cell and expelling others. It also pumps protons (H^+) out of the cytoplasm. In a living bacterium, these energy-consuming activities maintain an electrical gradient across the CM with the inside of the membrane being negative relative to the outside. The pH is also more acidic just outside the cell. The higher concentration of protons outside the CM relative to the cytoplasm creates the proton motive force[12] (PMF) that continually drives protons into the cell. As each proton crosses the membrane via an embedded protein complex, the cell transduces the electrical energy of the PMF into usable chemical energy in the form of ATP. This is an essential mode of cellular ATP production.

As virions are being assembled, the holins accumulate steadily in the CM with no apparent effect on virocell activity – no effect, that is, until their concentration reaches a critical threshold. Along the way, as their concentration in the CM increases, they aggregate into protein "rafts" in the membrane. As more holins enter the CM, the rafts grow larger until one of them reaches critical size. Then, suddenly, the holin timer goes off. The membrane, compromised by the accumulated holins, can no longer maintain the electrical gradient and the PMF. The clustered holins change conformation and construct openings in the CM. Some holins "trigger" to make small pores, while others form a huge gaping

---

[11] transmembrane domain: a protein domain with an α-helical secondary structure and containing many hydrophobic amino acids that readily embeds in lipophilic membranes.

[12] proton motive force (PMF): a force created when the metabolically-driven proton pump drives protons across the CM to the outside of the cell, thereby producing a higher proton concentration outside the cell. The difference in concentration creates the PMF that pushes protons back in. This push is harnessed by the cell to generate usable chemical energy in the form of ATP.

hole. Any size hole allows free passage of protons and ions, and thus spells cell death. But cell death alone does not immediately release the virions from entrapment.

Temperance: The Classic Model

As mentioned above, Temperance (λ) is one of the phages that accumulates active endolysins in the cytoplasm until it is ready to lyse the virocell. To unleash these enzymes, the phage must make a portal for them through the CM. A folded endolysin is a large molecule, far too large to pass through any of the normal membrane transport channels. What is needed to free the virions is a hole spacious enough to allow numerous folded endolysin proteins to diffuse out of the cell, attack the nearby peptidoglycan, and thereby quickly lyse the virocell.

Temperance and many other phages use both a holin and an antiholin to fashion such a hole in the CM. Both proteins are encoded by the same gene, both translated from the same mRNA – a well-known phage trick. The mRNA transcript of this gene encodes a protein containing 107 amino acids. However, there are two ribosome binding sites in that mRNA, thus two sites where translation can begin. When a ribosome starts translation at the upstream site, the protein product will be the 107 amino acid antiholin. But a ribosome can instead bind to the second site, thereby skipping the first two amino acids. This yields the 105 amino acid holin. Thus, these two counterbalancing proteins are identical except for the two extra amino acids at the N-terminal end of the antiholin. It is the N-terminus that normally extends outside the CM into the periplasm. Since ribosomes bind to the second site about twice as often as to the first, approximately two holins are made for every antiholin. Tweaking this ratio is one of several ways that Temperance can fine-tune its holin timer.

Temperance produces all the participating proteins – holins, antiholins, endolysins, and virion structural proteins – at top speed starting between eight and ten minutes after infection. Virions and endolysins accumulate in the cytoplasm, while the holins and antiholins are stationed on site in the CM. The holin waits with all three transmembrane domains embedded in the membrane and its N-terminal end extending to the outside (see Figure 67). The antiholin would do the same, but it can't quite succeed. The two extra amino acids at its N-terminal end include one that carries a positive charge. The charged state of the CM of a living cell prevents the antiholin's charged N-terminus from penetrating the membrane. This, in turn, restrains the adjacent domain from embedding. With this altered topology, the antiholin cannot act as a holin, and moreover its presence inhibits aggregation of the em-

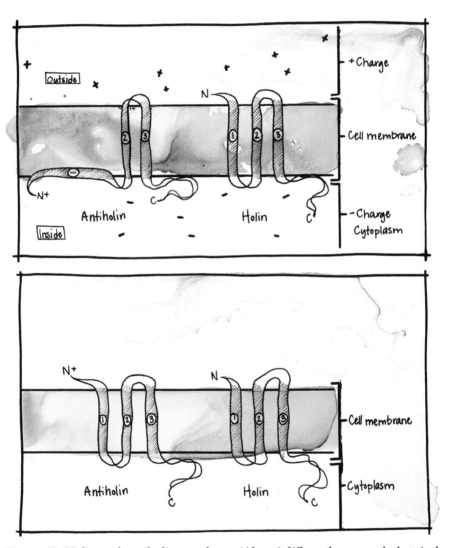

Figure 67: Holin and antiholin topology. (Above) When the normal electrical gradient is present across the membrane, all three segments of the holin protein embed in the CM, but the first segment of the antiholin remains outside the CM. (Below) When the membrane gradient is destroyed, the antiholins embed fully and function as holins.

bedded holins. Both holins and antiholins accumulate in the CM as dimers. Two holins together comprise a homodimer that can participate in hole formation. When a holin dimerizes instead with an antiholin, the resulting heterodimer is inactive.

Meanwhile, apparently unaware of the lit fuse, the virocell continues to swim along as it takes up nutrients and synthesizes phage DNA and proteins. Then, about 50 minutes post infection when 1000 to 3000 holins have accumulated, the concentration of holin homodimers in the CM reaches a critical threshold. At that point, some of them are triggered to form the first small gap in the membrane. This destroys the electrical gradient that had been preventing the antiholin from burying completely within the membrane. Suddenly, all the antiholins embed fully and now function as holins, instantly multiplying the concentration of active holins. The result is a huge hole in the CM – 340 nm across on average, with some giants extending as much as 1 µm. Picture such a hole made by Temperance in a rod-shaped *E. coli* that is ~0.5 µm wide and 2 µm long. Not surprisingly, virocell death is instantaneous. Almost as quickly, enough endolysins exit through the hole to effectively attack the nearby peptidoglycan. Within a minute, the weakened cell wall can no longer withstand the turgor pressure and the cell explodes.

If it is the holins that do the work of hole formation, why bother to make antiholins? Temperance and many of its relatives can lyse the cell just fine when experimentally deprived of their antiholins, albeit slightly earlier. For these phages, antiholins are a refinement that fine-tunes lysis timing. However, in some other phage families, the antiholin brake is essential for a productive infection. Without it, lysis occurs extremely early, before any virions have assembled.

Since *E. coli*, Temperance's host, is a Gram-negative bacterium, its cell wall is only a thin layer of peptidoglycan that is easily dealt with by a few active endolysin molecules. But outside this lies the cell's second membrane, the OM, which is not disrupted by either the holins or the endolysins. If Temperance relied on only those two agents, its virions would remain trapped inside the OM until the dead cell disintegrated. Spanins to the rescue! Genes for the spanins are often included in the lysis cassette, along with those for the endolysin and holin, in phages that infect Gram-negative hosts. While the holins were accumulating in the CM, two spanin proteins were also accumulating in the cell envelope. The "inner" spanin waited in the CM, the "outer" spanin in the OM. As their name suggests, together they span the periplasm to bridge the two membranes. Rope-like bundles containing eight to ten copies of each spanin thread their way individually through lacunae in the peptidoglycan. There they wait, each bundle isolated from its neighbors by the surrounding peptidoglycan strands until that peptidoglycan is cleared by the endolysins. After the endolysins have done their work, the neighboring bundles oligomerize[13] and fuse the two membranes to form large open channels that traverse them both. The virions are free!

Spanins are an essential part of the lysis equipment for Temperance and many other phages that use a holin/endolysin mechanism to lyse their Gram-negative virocells. Knowing that, you might well be wondering how Minimalist and other phages that use protein antibiotics[14] to lyse Gram-negative cells manage without them. Perhaps the answer lies in the covalent links that staple the peptidoglycan to the OM. Weak regions in the cell wall that result from inhibition of peptidoglycan synthesis would create zones where the OM is not as firmly secured to the peptidoglycan. Such zones, in turn, weaken the OM. Soon the defective cell wall together with the weakened OM can no longer contain the turgor pressure, and the resulting blow-out ruptures them both. Contrast this with the localized action of Temperance's endolysins. They digest a channel through the cell wall but the intact OM is capable of holding things together for a while.

Overall, Temperance's lysis strategy echoes some of the mechanisms for self-regulation seen in virion assembly. The needed lysis proteins are all synthesized concurrently throughout the 50 minute period of virion assembly. They accumulate silently, poised for action. When lysis begins, each protein does its job in turn. The correct sequence of actions is ensured because each step requires completion of an earlier

---

[13] oligomer: a complex formed by the association of a few macromolecules, typically proteins, usually non-covalently bound.

[14] Proteins that cause lysis by inhibiting peptidoglycan synthesis are referred to as protein antibiotics since their action mimics that of the β-lactam antibiotics, e.g., the penicillins and cephalosporins.

one. Hole formation by the holins requires prior holin synthesis and localization in the CM. Peptidoglycan digestion by the endolysins can occur only after the holins have done their work. Synthesis of virion components halts at the time of hole formation. Less than a minute later the virions are on their way – no point in wasting time hanging around in a corpse.

Tweaking the Holin Timer

The superior phage strategy would also set the lysis timer for the optimal time. Since conditions such as host availability are forever shifting, the expert phage would have a way to adjust the timer to lyse earlier or later. This means a way to adjust exactly when the holins trigger. That time, in turn, depends on both the rate at which the holins accumulate and the concentration of active holins required in the CM for triggering. Theoretically, Temperance could adjust either of these factors. In actuality, it synthesizes both holins and endolysins at full speed starting about eight minutes post infection and continuing until lysis. Even given that constraint, it could adjust lysis time by shifting the proportion of holins and antiholins made. That proportion depends on how often the ribosomes bind to the holin translation start site versus the antiholin site. Taking a lesson from Minimalist, one could imagine that Temperance might modify mRNA folding to adjust ribosome access. However, there is a difference here. Minimalist is regulating ribosome access to the translation start sites on two different genes (the maturation protein and the capsid protein), thus to two sites separated by hundreds of nucleotides. For Temperance, the competing ribosome binding sites are essentially adjacent, thus would be seemingly impossible to shield differentially.

Temperance's strategy is to tweak the holin concentration required for triggering in order to either move up or delay lysis. Such tweaking, in fact, can be readily done without preventing the holin from doing its basic job. A single amino acid substitution in critical regions within the holin protein can trigger lysis as early as 20 minutes post infection, delay lysis indefinitely, or set the timer for any time in between. An amino acid substitution requires only a single missense mutation. Such mutations occur due to replication errors, albeit at a low rate (see PIC). Nevertheless, in any natural environment where Temperance-like phages are abundant, there will be some phages with variant holins that trigger a bit earlier or later than others. A phage that lyses a tad earlier will prosper under conditions such as high host abundance. Conversely, when host availability changes, those that delay lysis slightly will have the competitive advantage and will increase in the population. In this way, a phage lineage can continually adapt its lysis time to closely track the ever-changing optimum.

Tweaking the Timer on the Fly

This mode of adaptation through mutation can benefit subsequent generations, but it has no effect on the current infection. Of the many phages that lyse by a holin/endolysin mechanism, only one group is known to have a way to adjust their lysis timer on the fly. That skilled group includes Lander (T4) and its kin, and even they can respond to only one environmental factor: host abundance. Their surveillance system continuously assesses the likelihood that their progeny would encounter a member of its host species upon release. How can a phage replicating inside a virocell "know" anything about the host population outside? They infer scarcity or abundance by monitoring the number of competing virions that are still drifting nearby in search of a host. When potential hosts are abundant relative to these virions, each virion will quickly bind to one of them. This leaves fewer virions on the prowl at any particular moment. Conversely, when potential hosts are in short supply, there will be homeless virions knocking more often on the door of the already infected virocells.

Lander monitors those knocks starting at only three minutes post infection. Each time a related phage[15] attempts to infect its virocell, Lander postpones lysis for up to ten minutes. If another virion adsorbs during those ten minutes, the delay timer is reset to ten minutes. Given enough infection attempts, lysis can be delayed for six hours or more with virion production continuing apace all the while. Meanwhile, outside the virocell the number of potential hosts and competing virions may rise or fall. When host availability increases sufficiently, attempted infections become too infrequent to maintain inhibition and Lander proceeds with lysis. With this system Lander wins two ways. A brief postponement until host numbers increase may provide im-

---

[15] any of the T-even coliphages (T2, T4, T6) or their close relatives.

proved infection prospects for its progeny virions. A longer postponement yields a larger burst size which improves the chances that at least
one virion will score whatever the host abundance happens to be at
that time.

Having many *E. coli* in the immediate vicinity is not the same thing as
having many available hosts. Unbeknownst to a wandering Lander
virion, some of those *E. coli* may be virocells with a related phage replicating inside. Lander can still adsorb and irreversibly commit to infecting that cell, but it will be rebuffed by the phage in residence through a
tactic of superinfection exclusion. The superinfecting phage rarely suc-
ceeds in getting its chromosome into the cell. Often the chromosome
is delivered into the periplasm, or it may never even get out its capsid.
Either way, this is the end of the game for that virion.

For Lander, the blocking of delivery of a superinfecting chromosome
and the postponement of lysis are related. Both result from Lander's
response to the delivery of DNA or an internal protein into the periplasm by the would-be invader. This response is mediated by Lander's holins and antiholins, both of which function a bit differently than
we saw for Temperance. Lander's holin has only one transmembrane
domain that embeds in the CM. Its large C-terminal domain projects
out from the CM into the periplasm. As was the case for Temperance,
these holins accumulate quietly until they reach their critical concentration, at which time they trigger and create holes in the membrane.
Lander, too, uses an antiholin to counter the holins. Like its holins,
each antiholin embeds in the CM with most of its length extending
into the periplasm. But these antiholins aren't simply slight variants
of the holins; they are unrelated proteins that behave quite differently.
It is their periplasmic domains that inhibit triggering by binding the
periplasmic domains of the nearby holins. Unlike the holins that accumulate in ever greater numbers in the CM during the infection, the
antiholins stay in the membrane only briefly. Each is soon inactivated
and expelled to the periplasm where it will be degraded by a protease
within a few minutes. By about 20 minutes post infection there are
more trigger-happy holins waiting in the CM than the small steady-
state population of antiholins can block. The holins trigger, the virocell
is dead, and lysis ensues.

The wisdom of Lander's unusual antiholin strategy becomes apparent when a related phage attempts to infect the virocell. The intrusion stabilizes the active antiholins in the CM so that they, too, accumulate and keep pace with the holins, thus blocking the trigger. But this postponement is temporary. Unless there is another infection attempt within ten minutes, the antiholins will be released from the CM and destroyed. Thus, the primary function of these antiholins appears to be to respond to knocks on the door by briefly postponing lysis.

Pinholins

Punching holes in the membrane kills the virocell, but does not rupture its cell wall nor let the waiting virions escape. The virions are trapped until the cell lyses. Therefore, immediately after lethal holin triggering, the expert phage puts its waiting endolysins to work to weaken the cell wall and induce lysis. As is typical, these experts have devised more than one way to get active endolysins on site quickly. Temperance accumulates them in an active form in the cytoplasm, then creates such a huge hole in the CM that enough endolysins quickly exit and degrade the peptidoglycan in the vicinity. Some other phages think it better to station the endolysins, inactive, on the outside of the CM in advance, then instantly activate them all. To do this, they include a specific sequence of amino acids at the N-terminal end of the endolysin that instructs the cell's secretory system to deliver the endolysins to the CM. There the inactive enzymes wait quietly, tethered to the outer face of the CM, while the phage's holins accumulate in the membrane. These phages use pinholins. As their name suggests, this family of holins makes "pinholes," rather than gaping holes, in the membrane. These pinholins aggregate into clusters, seven pinholin proteins per cluster. Each cluster forms a ring with a central channel that is less than 2 nm in diameter. This is not large enough for an endolysin molecule to escape, but larger holes aren't needed since the endolysins were parked outside the membrane. Pinholes that allow protons and ions to move through freely are sufficient. The electrical gradient across the membrane vanishes. The resultant rapid rise of the pH outside the CM releases and activates the endolysins. The cell wall is attacked and lysis follows quickly.

A Pyramid Builder

This reliance on the turgor pressure to rupture the cell wall seems, in the eyes of some phages, to be embarrassingly crude, to lack elegance or flair. Rather than explode the cell, messily spilling the contents, why not build a door and then open it wide for your progeny? And in that case, why not give that door a distinctive architecture? Pharaoh[16] builds pyramids that open to form classic virion escape hatches.

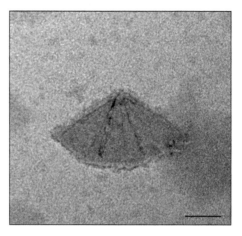

Figure 68: Classic architecture. Negative contrast TEM of a closed pyramid (side view) isolated from a *Sulfolobus* cell infected by the other known pharaoh, *Sulfolobus islandicus* rod-shaped virus 2 (SIRV2). Courtesy of Tessa E. F. Quax, University of Freiburg.

Pharaoh is an extremophilic archaeal phage that thrives, along with its crenarchaeal host, in the boiling acidic hot springs of Yellowstone National Park: temperature >75° C, pH <4. During an infection, it synthesizes one protein that strengthens and thickens small regions of the CM. In these locations it induces the formation of small, seven-sided pyramids–an unusual symmetry rarely seen in biological structures (see Figure 68). Numerous pyramids, each with sharply-defined facets, grow out-

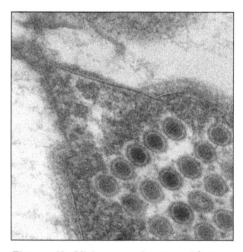

Figure 69: Virions await pyramid completion. Pyramid construction proceeds as Pharaoh's virions accumulate nearby. Courtesy of Sue K. Brumfield and Mark Young, both at Montana State University.

ward on the surface of the cell. That protein alone is both necessary

16 Pharaoh: *Sulfolobus* turreted icosahedral virus STIV, an unclassified extremophilic virus that infects *Sulfolobus*, a crenarchaeon

and sufficient to build pyramids—no other phage proteins required. When researchers modify *E. coli* to synthesize that one phage protein, pyramids form on its CM—even though the CM of Bacteria, including *E. coli*, is composed of different molecules than are found in the CM of all Archaea, including *Sulfolobus*.

As assembled virions accumulate during a Pharaoh infection, the pyramids grow to reach heights of 100 nm or more, and each side of the heptagonal base can be 180 nm. The CM of its *Sulfolobus* host is protected over the entire cell by a protein S-layer firmly anchored to the CM. Each emerging pyramid pushes a region of naked CM out through the S-layer. The pyramid interior is continuous with the virocell cytoplasm and is occupied by 50-150 densely-packed icosahedral virions awaiting release (see Figure 69). At some point the larger pyramids open, presumably in concert, each one opening like a flower with each petal (facet) remaining attached at the base. The virions exit leaving behind cell ghosts with an intact CM and S-layer, punctuated by discrete holes where the larger pyramids had stood.

Pharaoh is one of only two archaeal viruses known to build pyramids, the other one belonging to a different family.[17] Even Pharaoh's closest relative lacks the pyramid gene. As for the rest of the crenarchaeal viruses that thrive in very high temperatures and acidity, many are thought to not be lytic. Their progeny escape, but we don't yet know how.

Phages are killers, but they are not ax murderers, nor do they kill on a whim or for sport. Rather they skillfully lyse their virocell when necessary to let their progeny out. This lysis is as sophisticated and precisely orchestrated as takeover, assembly, and every other step in the lytic life cycle. Every move is optimized and finely tuned. Timing is especially critical, and must adapt to changing conditions. The very few phages that travel inside skinny virions slip out quietly while the virocell continues to produce more virions. If indeed practice makes

---

[17] SIRV2, *Sulfolobus islandicus* rod-shaped virus 2, is a member of the rod-shaped *Rudiviridae*.

perfect, phage perfection is not surprising. Phages have had an inconceivable amount of practice. A rough estimate is that phages launch 10^{23} successful lytic infections every second, or 10^{29} per year – a number roughly equal to the number of atoms in your body. And this has been going on for billions of years. Each lytic event sends 25 or 50 or more virions on a quest for a new host in which to repeat this cycle yet once again.

Further Reading

Bernhardt, TG, I-N Wang, DK Struck, R Young. 2002. Breaking free: "Protein antibiotics" and phage lysis. Res Microbiol 153:493-501.

Brumfield, SK, AC Ortmann, V Ruigrok, P Suci, T Douglas, MJ Young. 2009. Particle assembly and ultrastructural features associated with replication of the lytic archaeal virus *Sulfolobus* turreted icosahedral virus. J Virol 83:5964-5970.

Quax, TEF, S Lucas, J Reimann, G Pehau-Arnaudet, MC Prevost, P Forterre, SV Albers, D Prangishvili. 2011. Simple and elegant design of a virion egress structure in Archaea. Proc Natl Acad Sci USA 108:3354-3359.

Quemin, ER, TE Quax. 2015. Archaeal viruses at the cell envelope: Entry and egress. Front Microbiol 6:552.

White, R, S Chiba, T Pang, JS Dewey, CG Savva, A Holzenburg, K Pogliano, R Young. 2011. Holin triggering in real time. Proc Natl Acad Sci USA 108:798-803.

Young, R. 2013. Phage lysis: Do we have the hole story yet? Curr Opin Microbiol 16:790-797.

Young, R. 2014. Phage lysis: Three steps, three choices, one outcome. J Microbiol 52:243-258.

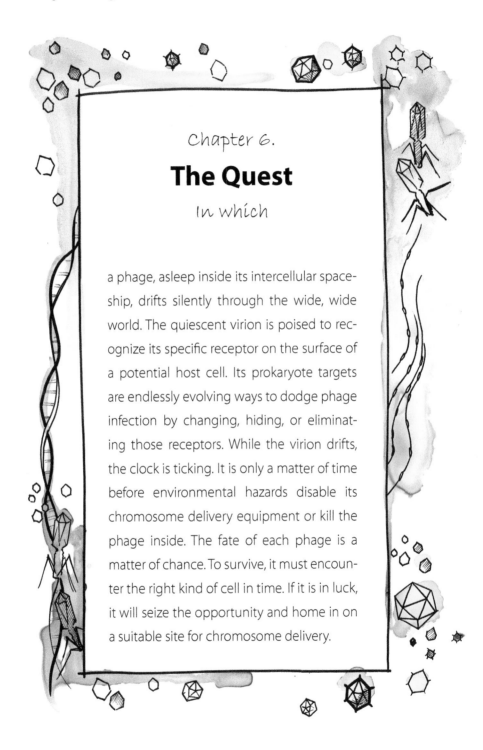

Chapter 6.

The Quest

In which

a phage, asleep inside its intercellular space-ship, drifts silently through the wide, wide world. The quiescent virion is poised to recognize its specific receptor on the surface of a potential host cell. Its prokaryote targets are endlessly evolving ways to dodge phage infection by changing, hiding, or eliminating those receptors. While the virion drifts, the clock is ticking. It is only a matter of time before environmental hazards disable its chromosome delivery equipment or kill the phage inside. The fate of each phage is a matter of chance. To survive, it must encounter the right kind of cell in time. If it is in luck, it will seize the opportunity and home in on a suitable site for chromosome delivery.

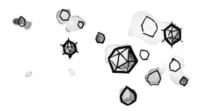

Not all those who wander are lost.
J.R.R. Tolkien, *The Fellowship of the Ring*

The true adventurer goes forth aimless and
uncalculating to meet and greet unknown fate.
O. Henry

Of course. Treasure hunts make much better
stories when there's treasure at the end.
Eric Berlin

Say it, reader. Say the word 'quest' out loud. It is
an extraordinary word, isn't it? So small and yet
so full of wonder, so full of hope.
Kate DiCamillo, *The Tale of Despereaux*

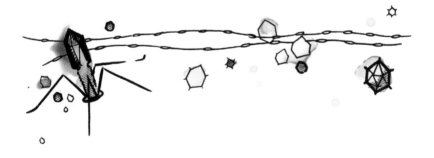

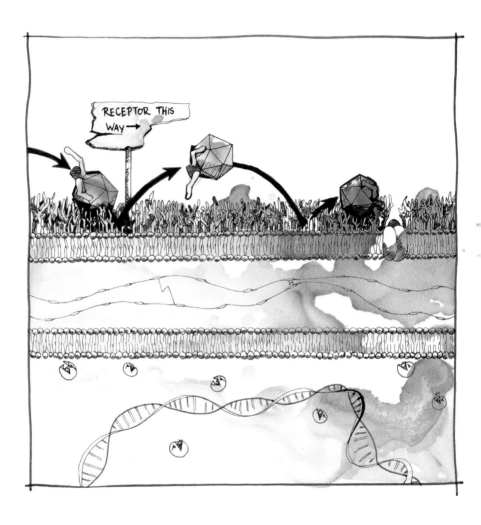

revious chapters portrayed our pheatured phages in action as they replicated and sent their progeny out into the world. When it has been packaged and released from the virocell, a phage is reduced to a dormant chromosome within a proteinaceous shell. Now it drifts like a seed or a spore, helpless and at the mercy of winds and currents. Like those more familiar agents of dispersal, a virion carries a packet of information that is capable of self-replication when provided with the right conditions. Given the opportunity, a phage can commandeer a prokaryotic cell and turn it into a virion factory. Not just any prokaryote will do – only a bacterium or archaeon of precisely the right kind. Moreover, the virion and its cargo must survive long enough to find that cell. If the virion meets a host in time, the phage may activate and multiply. Otherwise, it will perish. The odds are against success, just as they are for any individual plant seed. But, only one virion on average from every burst needs to succeed in order for the lineage to survive.

Hic Sunt Dracones

The journey is hazardous. Exposure to ultraviolet radiation (UV) or other DNA-altering agents can inflict lethal damage on a phage chromosome – serious damage such as chromosome breaks. UV-induced dimers[1] formed by the linkage of adjacent bases block subsequent DNA replication. Each hour, this form of death can befall up to 5% of the phages in transit in sunlit oceanic surface waters. Rescue is sometimes possible if the virion itself remains intact and is able to deliver the damaged chromosome into a host cell. There the phage has access to the cell's own DNA repair mechanisms. Bacteria[2] are, of necessity, equipped to cope with the same sorts of damage. They are able to excise dimers and replace them with the correct bases, ligate DNA strand breaks, and cobble together a functional chromosome by homologous recombination between two damaged strands. If a phage arrives in a cell that also suffered damage, the bacterium may already have acti-

---

[1] In DNA, dimers form from adjacent pyrimidine bases, those bases being T and C. Thus, dimers can be T-T, C-C, or T-C.

[2] Once again I refer to Bacteria, ignoring the Archaea, because so much more is known about the Bacteria and their phages.

vated its recombination machinery to resolve its own problems. Temperance (λ) relies mostly on its host for chromosome repair services, while Lander (T4), with its larger genome, encodes at least ten proteins to help with restoration. Phages such as Skinny (Ff) with ssDNA chromosomes are particularly vulnerable, because they don't have a second DNA strand to assist with repair.

In addition, recombination between two phage chromosomes, along with a bit of luck, can fortuitously increase the likelihood of phage survival. When two or more phages of the same type happen to be infecting the same cell, homologous recombination between their chromosomes can occur during their replication. If both phages carry lesions, recombination between them sometimes yields a lesion-free chromosome. Temperance has an additional strategy that Lander lacks: the bacterium may carry the chromosome of a related phage embedded in its own chromosome (see "Coalition" on page 241). This offers additional opportunities for recovery by homologous recombination.

Phage virions are outstanding transport containers. Just how outstanding was evident when curious researchers put the virions of one marine phage[3] through a comprehensive torture test. The evidence? Its virions were still able to infect a host after exposure to temperatures ranging from $-196°$ to $65°$ C, or after spending 24 hours in acid or alkaline conditions (pH 3 to 10), in seawater with more than twice the normal salt concentration, or under pressures equivalent to being 5000 m deep in the ocean. Capsids also protect the phage chromosome from attack by enzymes such as DNase and RNase. Nevertheless, some virions succumb en route. Virion life expectancy varies. Survival times differ in solar salterns and bubbling acidic hot springs, in glacial ice and on sunlit leaves, in the gut of an animal and in deep ocean sediments. In most situations we simply don't know how long virions can remain not only intact but also infectious. Their survival in the oceans is relatively brief, on the order of a few days. Here exposure to the UV in sunlight can damage the phage chromosome while leaving the virion structure intact. That structure, in turn, is under attack by the exoenzymes secreted by Bacteria to feed themselves.[4] Some protection

---

[3] *Vibrio* phage SIO-2.

[4] Bacteria use exoenzymes to predigest macromolecules to smaller molecules that can be imported through the CM.

against those enzymes is afforded by the close packing of the proteins in phage capsids, which makes the proteins less accessible to enzymatic attack. Capsid structure does not hide virions from hungry protists[5] that regard virions as nutritious snacks rich in carbon, nitrogen, and phosphorus. However, ingestion does not guarantee digestion. Even if "swallowed" and delivered into a protist "gut," some virions resist enzymatic digestion and are later excreted intact.

Strategic Hunting

Given the hazards en route, the skillful phage finds a host as quickly as possible. Success depends, in part, on host abundance. Even when Bacteria are numerous in the vicinity, potential hosts may still be scarce. A typical phage can infect and replicate within the members of only one bacterial species, often only one strain. If suitable hosts are rare, the probability of encountering one is low. If rare enough, the phage population declines and the few members of that prokaryote strain present escape phage predation. (This predator-prey dynamic has major implications for the maintenance of bacterial diversity, a story that will be developed in PIC.) But even when hosts are rare within the environment as a whole, they may be more abundant locally. Prokaryotes are clonal. When virions are released by virocell lysis, suitable hosts in the form of kin of the lysed cell are apt to be nearby, especially when they are mired together within bits of sticky, organic debris.

A dormant phage chromosome within a capsid appears helpless, buffeted by currents, dependent on random diffusion to carry it to a host. If it could navigate to locations where hosts are abundant, it would have greater success. Bacteria can sense a gradient of food and then actively move toward the source, but virions have no metabolic energy source and no means of locomotion. The best they can do is to seize the opportunities they encounter as they aimlessly wander. Or is it?

When a virion bumbles into good hunting territory, it would be wise to stick around long enough to encounter a host. Moreover, if these loitering virions release progeny in the same zone, the next generation begins their quest in a favorable locale. One such hunting ground is

---

[5] protist: an informal term for unicellular eukaryotes, an extremely diverse group of both autotrophs and heterotrophs that comprises ~20 phyla. Familiar protists include amoeba, paramecia, dinoflagellates, diatoms, and algae.

associated with animals. Vulnerable exterior surfaces ranging from the exterior of coral polyps to the lining of our own gut and lungs are covered with a protective layer of mucus. Compared to the surrounding milieu, these layers house greater concentrations of Bacteria. Some of these Bacteria are pathogens looking for a way into the animal's tissue, but more are members of diverse communities of beneficial microbes.

Mucus gets its characteristic viscous, gooey nature from its network of mucin glycoproteins.[6] To a bacterium, mucins offer a hospitable bed and breakfast: a bed formed by the mucin mesh, a breakfast composed of the mucin-linked glycans. Bacteria that take advantage of these accommodations don't escape phage attack. The channels in between mucin strands are spacious enough to allow the passage of nanometer-scale virions. Mucus is continually secreted by the underlying animal epithelial cells. The newer mucus pushes the older overlying mucus away from the epithelium and is, in turn, soon displaced by the ongoing secretion below. Eventually the oldest mucus sloughs off. A mucin protein may complete the journey from secretion to sloughing in a matter of hours or days. Bacteria counter this outward flux by swimming against the current toward the epithelium.

In order to stay around long enough to profit from the local host abundance, virions must counter the perpetual outward movement of the mucus conveyor belt. Without such tactics, the virions would be dumped into the environment along with the old mucus residues. Simply diffusing in all directions helps, but that alone isn't sufficient. Lytic phages can hitch a ride inside their virocell while they replicate. Since the virocell may continue swimming up to a minute before lysis, the progeny are often released deep within the mucus layer. (A temperate phage can fare even better, being onboard for longer. See "Till Death Do Us Part" on page 251.) Lander has its own specialized gambit that enables it to hunt more efficiently. Its capsid is decorated with 155 proteins[7] that project individually outward from the capsid surface. These proteins bind transiently to the mucin glycans and briefly arrest virion diffusion. Decorated with these proteins, Lander virions spend more time attached to mucin fibers, explore a smaller local region more thor-

---

[6] glycoprotein: a protein with attached glycans.
[7] Hoc protein (highly immunogenic outer capsid protein): a non-essential structural protein attached to the capsid of Lander (phage T4).

oughly, and encounter potential hosts more often. Many other phages have similar proteins somewhere in their virion that may confer comparable benefits in other environments, as well.

To Know a Host When You Meet One

In its erratic travels, a virion bumps into all manner of things – rocks, organic debris, protists, diverse prokaryotes. How does it distinguish a potential host from everything else? It detects a particular surface structure that it recognizes and can then bind. In other words, it looks for its receptor. Objects it encounters thus fall into one of two categories: potential hosts and everything else (see Figure 70). Any structure exposed on the outside of a bacterium is likely to serve

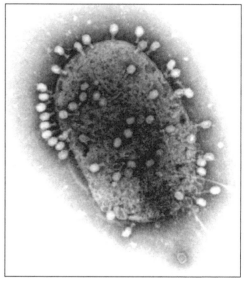

Figure 70: Lander (T4) virions have landed on the surface of an *E. coli* bacterium. Courtesy of Dr. M.V. Parthasarathy.

as the receptor for some phage or other. A receptor may be present at all times or it may vanish under some conditions. Similarly, a receptor may be relatively abundant and widely distributed over the cell, or it may be localized. Often it is located far from the phage's ultimate entry point on the CM.

Drawings of bacterial cells give the impression that they offer a smooth landing surface to passing virions: so smooth that a virion might be imagined to slide off, so smooth that the search for a receptor might be likened to a stroll along neighborhood streets in search of a house of a particular architectural style. But a virion quickly finds itself facing a tangled macromolecular jungle. What is a virion apt to encounter as it approaches a prokaryote cell? The outermost layer of some Bacteria is a thick overcoat of polysaccharides[8] that creates a capsule or slime

---

[8] polysaccharide: a polymer of saccharides, i.e., of sugars.

layer surrounding the cell wall and membranes. Capsules stump some phages by concealing the phage's receptor, but other phages use the capsule as their receptor and then work their way in from there. Moving inward, the surface of some Bacteria and Archaea is protected by a precise lattice formed by a monolayer of identical proteins–the S-layer. This is a double-edged sword for the cell. It deters some phages, perhaps by the negative charge on its surface or by the small pores (maximum 13 Å or 1.3 nm) in the protein mesh. Other phages rely on the S-layer to provide their receptor.

Numerous Bacteria and Archaea have cellular appendages that extend through the CM and any surrounding layers. The best known of these include the flagella and pili. Flagella are long, rotating, propeller-like filaments that move the cell rapidly through liquid. Pili comprise a diverse collection of transitory appendages that serve a variety of functions. Some form bridges between bacteria for the transfer of DNA from cell-to-cell; others are extended to attach to the substrate a small distance ahead, then retracted to pull the cell forward; and short ones enable the cell to adhere to nearby cells and other surfaces. All are used as landing points or receptors by some phages (see "Exploiting Cellular Appendages" on page 203).

Bordering the cytoplasm is the cell envelope composed of the CM and cell wall in all Bacteria, with the addition of the OM in Gram-negative Bacteria. An exposed OM offers a smorgasbord of potential receptors: lipopolysaccharides (LPS), porins, transport proteins, and other embedded or anchored proteins. For Gram-positive Bacteria, the exposed surface is composed mostly of the cell wall teichoic acids, specific components of which are often used as receptors. In short, whatever is exposed to the world is fair game for a phage.

Conversely, every phage has tailored some protruding virion component for receptor recognition, and some specialized structures have been added for just that purpose. Many phages with long-tailed virions use their long tail fibers (LTFs), although one eccentric group uses a unique filament at the top of its head instead (see "Exploiting Cellular Appendages" on page 203). Some with short tails employ tailspikes, while tailless icosahedral virions may rely on spikes stationed at each of the twelve vertices. Fiber, filament, or spike – all of these structures contain a receptor-binding protein (RBP)[9] positioned with its binding surface exposed and poised to interact with a specific receptor.

Commitment

Sometimes the phage adsorbs irreversibly to this initial receptor, thus triggering chromosome ejection. In this case, the phage has committed to infection of this cell and its quest is over. The site of irreversible adsorption is also the site of chromosome delivery. However, often adsorption is instead a two-step process. For these phages, first contact with the primary receptor establishes a weak and reversible bond lasting perhaps only a few seconds. Although brief, this is long enough to encourage the virion to end its random search through vast three-dimensional space and now focus its exploration on the two-dimensional cell surface. This shift greatly increases the likelihood of locating its secondary receptor, the actual site of chromosome delivery. Sometimes the primary receptor is far from the CM and the virion must penetrate deeper into the cell envelope to locate its secondary receptor. Binding to this site is the irreversible step that prompts chromosome delivery.

The overall success rate for a phage strongly depends on the frequency of collision with potential hosts, thus on host abundance. Although this factor is mostly outside the phage's control, there are tactics (see "Strategic Hunting" on page 190) that can help position the phage in more favorable locales. In addition, a phage can increase the number of potential hosts by being less picky. Although most phages thrive with efficient infection of one very specific host, some have traded the benefits of specialization for a broader host range. There are also trade-offs in the choice of receptor. Some receptors are restricted to a small region of the cell surface, requiring a search mechanism to zero in on them after the initial landing, while others are widely distributed and abundant. Also finding a member of the right bacterial strain may not be sufficient. Some receptors are present during only some stages in the cell's life cycle or under only particular environmental conditions (see "Receptors Come and Receptors Go" on page 209).

---

[9] receptor binding protein: the virion structural protein that recognizes and binds to the specific structure that serves as the phage's receptor on the surface of a potential host cell.

A second factor influencing a phage's success rate is how skillfully its virions turn a random collision into irreversible adsorption. If it is the capsid of a long-tailed phage that bumps into the cell first, the phage's RBP at the end of a tail fiber may be far from the receptor. There is no guarantee that the virion will collide in any particular orientation, but electrostatic charges on both capsids and cell surfaces may possibly favor a particularly advantageous alignment.

Throughout this process the clock is ticking. Once chance has brought you to the doorstep, every minute spent strolling about on the cell surface looking for a receptor is a minute wasted. The sooner you find the door and go in, the sooner your progeny will go out.

Dodges

Prokaryote cells have numerous ways to destroy an invading chromosome after it has arrived in the cytoplasm (see "Survival on Arrival" on page 61). Relying on a quick strike post entry is not the safest strategy – better to never let the wolf in the door in the first place. One way to block entry is to conceal the receptor. Receptors are located where they are in the cell in order to perform a needed function. Nevertheless, sometimes one can be hidden without compromising its usefulness by tucking the small segment that the phage recognizes out of sight. Another frequently used defense is to alter the recognized part of the receptor. Changing just one amino acid in a critical location may be sufficient to create a mismatch between the receptor and the phage's RBP. This defense has its costs. Such a change to the receptor may slightly reduce the cell's fitness. Prokaryotes, like their phages, are competing intensely with one another for who can produce the most progeny over the long term and thus come to dominate the population in the future. Even a minor change that reduces fitness by less than a percent matters. Such a change may be maintained in the population as long as phage predation exacts a higher price. While this tactic might fool the phage and give the cell a respite, that reprieve will be short-lived as the phage will quickly counter with a corresponding modification in its RBP. The genes encoding the RBPs are the fastest evolving genes in a phage genome. (See PIC for a discussion of significant ecological and evolutionary effects of this ongoing interplay.)

Scouting on the OM with a Short Tail

Podovirus Stubby (T7) is one of the few phages whose adsorption process has been observed closely. Still, even here, many questions remain to be explored. True to its name, Stubby has only a short tail. Attached to that are six LTFs each with an RBP at the tip. The LTFs are fragile, each being merely three identical proteins that run side by side the full length of the fiber. While cruising, Stubby raises and lowers each LTF in turn, keeping four or five of them safely folded back against its capsid. Holding them close protects them from damage and makes for a more compact, faster diffusing virion. Even with just one LTF lowered, Stubby still surveys a large volume. It drifts in search of a cell that bears its primary receptor, part of the LPS that is abundant in the OM of its Gram-negative *E. coli* host. When a lowered LTF contacts a receptor, it binds briefly. Before it lets go, an adjacent LTF is lowered. This LTF is apt to find a receptor nearby, in which case it, likewise, binds momentarily. Picture a virion that never lets go as it "walks" across the OM on its six LTF legs. We think Stubby does it this way, but it is possible that each LTF could let go before the next takes hold. In that case, the virion would "hop" across the surface, but with a greater risk of being swept away. Walk or hop, Stubby's three-dimensional search has now narrowed to an exploration of the cell's two-dimensional surface.

This receptor scouting continues until the tip of the tail contacts and binds a secondary receptor, also located in the OM, that is Stubby's ultimate destination. Then walking halts and all six LTFs bind to the OM. In preparation for chromosome delivery, Stubby ejects its internal core proteins that assemble a conduit for DNA passage into the cell (see "Delivery with a Stubby Tail" on page 227).

Stubby Scores

An animation, publicly available online (http://bit.ly/1Tk7WdU), portrays Stubby (phage T7) as it contacts, recognizes, and adsorbs to a host, and then prepares to deliver its chromosome. Courtesy of B Hu, W Margolin, IJ Molineux, and J Liu, University of Texas at Austin and the University of Texas Health Science Center at Houston.

For Stubby, adsorption takes place entirely at the OM. When confronting an *E. coli* strain such as K1 that surrounds its OM with a polysaccharide capsule, Stubby is stumped. Other phages are not deterred. They nibble their way through the capsule, forming a tunnel that also protects the virion from being set adrift again by currents or collisions. These phages can further extend their host range by equipping their virion with multiple enzymes, each able to digest a different capsule type.

Like Stubby, Chimera (P22) has a short tail and infects a similar Gram-negative host (*Salmonella*), but unlike Stubby it has no LTFs (see Figure 71). Instead its tail ends in a long (24 nm) needle-like structure surrounded by six thick, short, immovable tailspikes. All of the tailspikes bind to receptors and never let go.

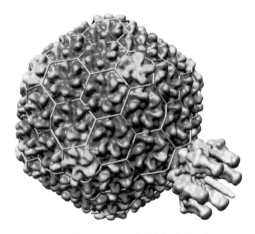

Their receptors are the long polysaccharide chains of the LPS that project outward from the cell's OM, a factor that puts Chimera's

Figure 71: Chimera's (P22's) virion has no tail fibers, but relies instead on six immovable tailspikes that surround its puncturing and delivery device. Courtesy Gabriel Lander, The Scripps Research Institute.

landing platform farther from the CM. The tailspike proteins are also enzymes that nibble away at the polysaccharides. Bite by bite, the virion eats its way closer to the CM, tail first. When the needle touches the CM, contact triggers chromosome delivery.

Phages can increase the number of strains they can infect by carrying multiple RBPs and multiple enzymes able to digest a pathway to their entry site. How many of each a phage can add is limited by genome size and virion architecture. If the phage is restricted to one RBP or enzyme, it can substitute a different option by swapping RBPs or enzymatic domains with other phages. Phage evolution is so rapid that such exchanges are a significant mechanism of phage adaptation (see

PIC). One phage in particular,[10] a large myovirus, warrants mention here for its unusually broad host range. It infects *E. coli* strains with and without capsules, as well as many strains of *E. coli*'s cousin, *Salmonella*. Its virions, fitted out with three different tail fibers and multiple spikes bearing at least five different RBPs, are the prototype of the Swiss army knife.

Landing on the OM with a Long Tail

Lander's tail and tail fibers are both much longer than Stubby's, but both phages use some of the same tactics when adsorbing to their *E. coli* host. Like Stubby, Lander holds most of its LTFs close to its capsid while on its quest (see Figure 72). Three or four of them are usually folded back with each held in place by a whisker, the whisker collar, and a capsid vertex. As one is raised, another is lowered – a movement that does not require an energy input.[11] Adopting this compact profile makes for more rapid diffusion and protects the spindly LTFs from damage, even while the LTFs continue to scan a large area. Lander can also retract all of its LTFs simultaneously and hold them all close so that they form a "jacket" surrounding its tail. However, in this configura-

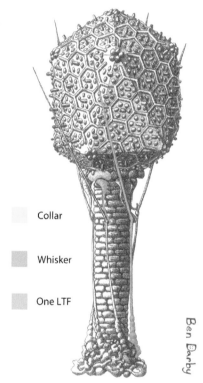

Collar

Whisker

One LTF

Ben Darby

Figure 72: The virion of Lander's very close relative, RB69, plays its cards close. Original drawing by Ben Darby. Previously published in *Life in Our Phage World* by Rohwer, F, M Youle, H Maughan, N Hisakawa. 2014. Wholon. Used with permission.

tion Lander's virions are not infectious. Lander resorts to this extreme measure when conditions are not favorable for infection. Its preferred

---

[10] phage φ92

[11] Four movies are available here http://bit.ly/2aM8khC provided you have e-journal access to PNAS. They present three-dimensional tomograms of Lander virions showing some LTFs extended and bound to a cell, while others are retracted. See Hu et al. 2015 in "Further Reading" on page 212.

habitat (the gut of humans and other mammals) is warm and approximately neutral pH. Lander, along with its *E. coli* hosts, is subject to occasional expulsion from Eden. When the temperature drops well below 20° C, it retracts all LTFs for their protection and to defer infection. A pH of 5 or less, which Lander encounters in the stomach on its way to the colon, prompts the same defensive response. Thus, in Lander's hands (or whiskers), LTFs function as both environmental sensors and adsorption aids.

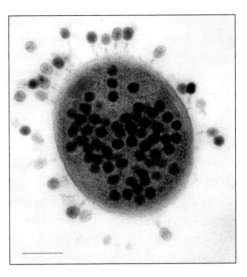

Figure 73: An *E. coli* cell, late in a Lander infection, with assembled virions inside. Its surface is peppered with adsorbed Lander virions, some of which (the empty capsids) have ejected their DNA. Bar = 300 nm. Courtesy of John Wertz, Yale University.

Some strains of *E. coli* carry about 10^5 copies of Lander's receptor[12] distributed over the cell's OM, and in other strains Lander uses a component of the LPS. In either case, locating one on the surface doesn't require much hunting (see Figure 73). However, Lander prefers a location near a cell pole for irreversible adsorption, the site of infection. So, like Stubby, it "walks" across the surface, relying on reversible adsorption to tether the virion to the cell. One extended LTF binds temporarily, and before it is released, a nearby LTF is extended, finds another receptor, and binds. Step-by-step, hand-over-hand, Lander explores the two-dimensional surface until an LTF binds a receptor in a favorable location. Then the LTF lingers while additional LTFs are lowered and also attach. As more LTFs bind, Lander's six short tail fibers (STFs) are freed from the tail baseplate and swing down to irreversibly adsorb to their LPS receptors. At this point, the virion is committed and poised for the dramatic structural changes that then launch an infection (see "Delivery by Lander" on page 226).

---

[12] OmpC, outer membrane porin protein C.

Potential receptors in the OM also include the porins. Heterotrophic Bacteria such as *E. coli* are opportunists that can utilize any of a number of sugars as their carbon and energy source. To efficiently scavenge glucose, galactose, lactose, sucrose, and many other sugars from the environment, they have dedicated porins and other transporters in their membranes. The phages, in turn, exploit these transporters for their adsorption or entry. For example, the OM porin protein (maltoporin) that actively imports the disaccharide maltose into *E. coli*, serves also as Temperance's (λ's) primary receptor. When there is no maltose in its environment, *E. coli* conserves resources by making little maltoporin. When maltose is present and being used, each cell may have 10,000 of these channels in its OM arranged in a spiral pattern spanning the length of the cell. In this situation, it doesn't take Temperance long to locate one. But adsorption is a two-step process for Temperance. For step 2 – irreversible adsorption and chromosome delivery – Temperance needs to locate its secondary receptor[13] in the CM. This receptor is also a component of the CM maltose transporter, but unlike maltoporin, it is localized near the cell poles. Thus, no matter where a virion first lands, it must make its way to a polar region. Temperance manages to do this without losing its grip by hitching a ride on the maltoporin. The majority of these proteins are mobile within the OM and migrate toward a pole at speeds sufficient to take them there from any location within a few minutes. This delivers Temperance to its secondary receptor, at which point the virion binds irreversibly and proceeds with infection.

Landing on the OM Without a Tail

Neither tailspikes nor LTFs – not even a tail – is a required accoutrement when infecting *E. coli*. Consider Yoda (ϕX174). Its virion is a small icosahedron adorned only with a spike at each of the twelve vertices. Each spike assembles from two proteins. Five copies of the larger of the two proteins are arranged in a ring that forms a spike structure with a potential DNA delivery channel through the center. Yoda's minimal spike extends only 3.2 nm above the surface of the capsid – quite a contrast with the long reach of Lander's 144 nm LTFs.

---

[13] ManY: a protein that is part of the CM mannose transporter in *E. coli* that serves as the secondary receptor for Temperance.

One copy of the smaller protein, the "pilot" protein, functions during DNA delivery (see "Yoda's Grand Entry" on page 229). Initially the capsid protein itself adsorbs reversibly to the polysaccharide chains of the abundant LPS in the OM. Subsequently, both spike proteins interact with Yoda's not-yet-identified secondary receptor[14] to launch the infection. One can imagine that after colliding with a potential host, a short roll on the OM surface would be sufficient to bring one of the spike-bearing vertices into contact with a secondary receptor.

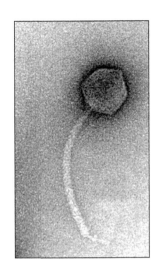

Figure 74: Positivist (SPP1) makes do with a single tail fiber. Courtesy of Rudi Lurz, Max-Planck Institut for Molecular Genetics, Berlin, Germany and Paulo Tavares, Department of Virology, I2BC, Gif-sur-Yvette, France.

Penetrating the Great Peptidoglycan Wall

Phages that infect *Bacillus subtilis* or other Gram-positive bacteria touch down on a landscape of cell wall peptidoglycan and teichoic acids (see "Gram-Negative, Gram-Positive, or None of the Above" on page 163). Both of these components have been exploited as receptors by various phages. Positivist,[15] for example, recognizes the abundant teichoic acids on the surface and briefly attaches to them with the tip of its long, flexible tail (see Figure 74). From this vantage point, temporarily tethered to the cell, it scans the area in search of its secondary receptor,[16] a protein embedded in the CM in the polar regions. At first glance it seems more challenging to find your receptor if it is restricted to only two regions of the cell surface, but this receptor waves a flag that makes it easier to locate. While much of this protein weaves back and forth through the membrane, one end is free to extend outside the cell. This end is long enough to reach through the thick cell wall – up to 55 nm – and beyond. Because of this flag, a virion does not have to penetrate the formidable peptidoglycan barrier

---

[14] Yoda may have two separate receptors, or it may use two parts of the same receptor molecule. See Michel et al. 2010 in "Further Reading" on page 212.

[15] Positivist: SPP1, a siphovirus that infects Gram-positive *Bacillus subtilis*.

[16] YueB: the CM protein of unknown function that serves as SPP1's receptor. Homologs of YueB are widespread among the Gram-positive Bacteria.

Figure 75: Nibbling your way through. Dynamo's short tail can't reach all the way through the thick cell wall of its Gram-positive host, *B. subtilis*. To reach the CM, the knob at the end of that short tail enzymatically digests a corridor for virion passage. This process is facilitated by the initial binding of the tailspikes to the ends of the teichoic acids that extend outside the cell wall. They, too, have enzymatic activity. They nibble at the free end of the teichoic acid chains, thereby shortening them and pulling the virion closer to the cell surface, perfectly positioned for DNA delivery.

in order to contact its receptor in the CM. A virion that finds its way to a polar region can readily contact this receptor, then bind irreversibly and launch an infection.

Dynamo (ϕ29), on the other hand, meets the same Gram-positive challenge equipped with a short non-contractile tail – a mere 38 nm long. But it also has, and uses here, 55 head fibers, 12 tailspikes, and a knob at the end of its tail. When Dynamo collides with a potential host, the head fibers, tailspikes, and tail knob can all contact the cell surface (see Figure 75). The unusual head fibers each project from a facet of the prolate capsid. Although not required for infection, they assist by keeping the virion from drifting away and aligning it perpendicular to the cell surface. The tailspikes, attached at the "neck" of the virion, fluctuate between two conformations with about half being "up" and half being "down" at any time. They are equipped with an enzyme that shortens the teichoic acids anchored in the cell wall by repeatedly cropping their free ends. Nibble by nibble, the virion is pulled in closer and the tail knob is brought int contact with the cell wall peptidoglycan. The tail knob also has an enzymatic activity, one that digests the peptidoglycan and clears a path for the virion all the way to the CM.

Exploiting Cellular Appendages

All of the phages we've watched so far go expertly about their stealthy business, albeit rather modestly. If you are looking for adsorption with more flair, consider Cowboy (χ).[17] This phage lassos a rotating flagellum and then rides it to the cell surface, landing close to its receptor (see Figure 76). Its prey, *E. coli*, swims rapidly using flagella that project through the cell envelope at many locations. All the flagella rotate in the same direction at more than 100 rpm. Moreover, they all simultaneously reverse their direction every second or two. When all are "rowing" counterclockwise, the cell "runs" forward at 20 µm per second, a distance equivalent to an impressive ten cell lengths every second. (The comparable speed for a human with a six foot "cell length" would be 24 miles per hour. Exceptionally fast humans clock a four minute mile, i.e., a speed of 15 miles per hour.) When *E. coli's* flagella all rotate in reverse, the cell tumbles and changes direction. Any way you look at it, flagella present a moving target for Cowboy.

For a lasso, Cowboy has a single long tail fiber that curls into a coil near its tip. Each flagellum is composed of many identical flagellin proteins stacked in a precise helical array, a pattern reminiscent of long phage tails. The resultant helical grooves turn a flagellum, in effect, into a threaded bolt. Picture the end of Cowboy's tail as a right-handed nut that can screw onto the flagellar bolt when the flagellum rotates counterclockwise. Switch to clockwise rotation and the nut spins off the bolt. When Cowboy lassos a flagellum, will it ride to its receptor or be sent flying off the end of the flagellum? It depends on where it attaches—close to the cell or farther away. Cowboy's lasso can travel a few micrometers along the flagellum per second. The flagellum might be six or seven micrometers long, and a run can last up to two seconds. Thus, Cowboy doesn't make it all the way to its target every time, but it does so often enough. When presented with an *E. coli* without actively rotating flagella to harness, Cowboy's rate of infection is undetectable.

---

[17] *Salmonella* phage χ

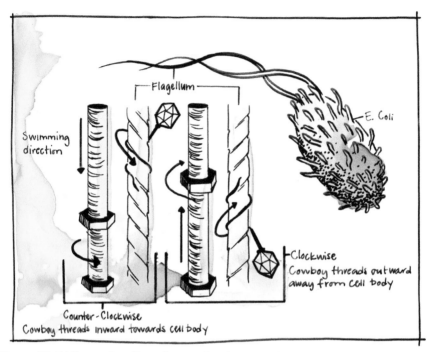

Figure 76: Riding a flagellum home. Cowboy's curly tail fiber wraps around a rotating flagellum, like a nut threading onto a bolt. Counterclockwise rotation spins the virion toward the cell surface, whereas clockwise rotation sends it in the opposite direction.

There are other caveats for flagellotropic[18] phages. Successfully zooming to the cell surface doesn't necessarily mean Cowboy has reached home. Similar flagella built from similar flagellins and with similar surface grooves are found on many different kinds of bacteria. A virion might ride a flagellum to the cell surface only to find it has arrived at the wrong door. In addition, if you require flagella for infection, you are up a creek when your host foregoes their production, which some bacteria do when starved. Without flagella, *E. coli* escapes infection by Cowboy. Not a big loss for Cowboy, you might think, because infection of a starved cell will yield few, if any, progeny. Better to pass the cell by and take your chances on drifting into an area with better nourished hosts. On the other hand, flagella aren't always an indicator of a well-fed bacterium. Some soil dwelling species[19] produce flagella when food is scarce, presumably to enable migration to a bet-

---

[18] flagellotropic
[19] For example, *Pseudomonas fluorescens*.

ter location. There is no perfect tactic that works all the time for every phage.

Phages know more than one way to lasso a rotating flagellum. For instance, Nerd (ϕCbK)[20] has a long tail available, but it uses its head (capsid) instead. For its lasso, it uses the long (200 nm), coiled filament that projects from the capsid vertex directly opposite the tail (see Figure 77). Its target is the single flagellum that extends from one pole of its host, *Caulobacter crescentus*. Clockwise flagellar rotation propels the cell forward with the

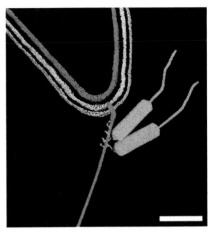

Figure 77: Nerd captures a host. Cryo-EM tomography segmentation showing Nerd with its head filament wrapped around a *Caulobacter crescentus* flagellum. Courtesy of Rebecca S. Dillard, Wright Lab, Emory University.

flagellum at the stern, but Nerd makes its move during the periods of counterclockwise rotation when the cell swims "backwards." Then, the virion threads along the flagellum toward the cell. Nerd still needs to bring its tail fibers into contact with its receptors located near the base of the flagellum. Picture a virion hanging from a flagellum by its head filament as the cell swims backwards, flagellum first. Resistance from the water sweeps the virion tail towards the cell, which positions the tail fibers close to the base of the flagellum. Nerd's essential receptors are in the CM nearby. Without its head filament, Nerd can still launch an infection, but its receptor search is too inefficient to compete successfully in the real world.

A Puzzle

Head filaments are rare, having been found in only a handful of different phages so far. Their assembly remains particularly puzzling. How does the assembling virion know which vertex should get the filament? In comparison, designating a single vertex to serve as the portal and tail attachment site is simple. As

[20] Siphophages ϕCb13, ϕCbK, and their kin

observed in Chimera (P22), for example, the portal assembles first. The scaffolding proteins then oversee the sequential addition of capsid proteins to construct the rest of the procapsid. To designate the vertex opposite the portal for head filament attachment implies having a way to identify that particular vertex. That vertex is distinguished by being the point on the capsid farthest away from the portal. Long-tailed phages use a tape measure protein to ensure precise, consistent tail length. Perhaps analogous measuring tapes extend out from the portal along the meridians. They would meet at the vertex farthest from the portal – the correct site for fiber attachment. Alternatively, if the procapsid grows outward in all directions from the portal, closure of the structure at that vertex might trigger filament attachment. If instead the filament, like LTFs, is added long after the procapsid is complete and scaffolding proteins have been dismissed, then we need some different thoughts.

Bacterial Conjugation

Bacterial conjugation[21] is the intentional transfer of DNA from one bacterium to another orchestrated by a conjugative plasmid. Conjugation is the way that the "selfish" plasmid is transferred to new cells that don't already have a copy, and thus spreads through a population. This transfer of DNA requires cell-cell contact. In *E. coli*, for example, some cells grow a sex pilus to serve as the conduit for DNA transfer between two cells.[22] Although the pili are part of the cell's structure, their production requires the presence and participation of a conjugative plasmid, of which there are many diverse types. A well-known one carried by some *E. coli* is the F-plasmid (F for Fertility). The plasmid encodes the proteins needed to induce and carry out pi-

[21] conjugation: a form of bacterial sex in which DNA is transferred from one bacterium to another through cell-to-cell contact. Often the process is coordinated by a conjugative plasmid that encodes the necessary proteins.

[22] Although it is clear that the conjugation pilus plays an essential role in bringing the conjugating Bacteria into close contact, some researchers maintain that the DNA is not transferred through the pilus channel, but by some other mechanism.

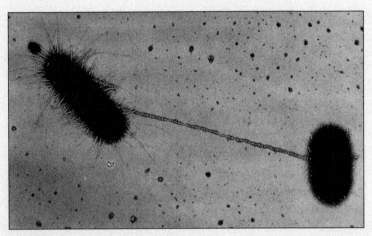

Figure 78: Unsafe sex. During conjugation in *E. coli*, the donor and recipient cells are connected by a sex pilus that provides the receptor for Minimalist and its kin. The hairlike projections especially apparent on the donor cell are fimbriae, i.e., short pili typically used to adhere to a substrate or to other cells. Courtesy of Charles C. Brinton, Jr.

lus production and DNA transfer. A pilus can grow up to 20 μm in length. Although these filaments are only ~9 nm in diameter, running the length of the filament is a central channel ~2 nm in diameter – wide enough to allow passage of ssDNA or ssRNA.

Pili are dynamic structures that go through alternate periods of extension and retraction by the addition or subtraction of component proteins at the base of the pilus. Cells containing the plasmid can have multiple pili growing from their surface at one time. If, during extension, the adhesive tip contacts another cell and adheres, the next retraction phase will bring the two cells close together (see Figure 78). Since the F-plasmid benefits by spreading through the population, its pilus wastes no time courting unacceptable mates. A sex pilus adheres only to cells that lack a copy, thereby ensuring that it accepts only cells without the plasmid as a mating partner. If the pilus adheres and retrieves a potential recipient, a copy of the plasmid DNA is then ported over to the recipient cell, thereby converting the recipient into a donor. The F-plasmid is referred to as an episome because it also occasionally inserts into the bacterial DNA. In this case a copy of the entire bacterial chromosome can be delivered, a process that takes about 100 minutes in *E. coli*. Although infrequent, this horizontal transfer of genes between Bacteria, both conspecifics and others, is an important mode of sexual recombination for bacterial populations.

Like other bacterial surface structures, the sex pilus of *E. coli* is exploited as a receptor by some phages. During the elongation phase, a sex pilus may extend several cell lengths and thereby serve as an inviting, sizable target for passing virions. All phages (so far) with ssRNA chromosomes adsorb to pili; co-liphages such as Minimalist

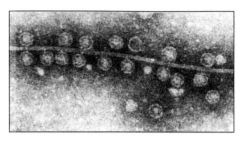

Figure 79: Still life with virions and pilus. A TEM of numerous Minimalist virions adsorbed to the outside of an *E. coli* sex pilus. Courtesy of Graham Beards, Wikimedia Commons.

(Qβ) use the sex pili of their *E. coli* host (see Figure 79). The hollow central channel within the pilus provides a conduit for ssRNA passage, but how the RNA exits the capsid and enters the pilus is not known. Shy (φ6), with its three dsRNA chromosomes, also uses the pili of its host to assist with its unusual mode of host entry (see "Shy's Delivery" on page 232).

First Hints for the Crenarchaeal Phages

Do the archaeal phages use the same collection of mechanisms as their bacteriophage counterparts? Even though we know little about them so far, it is safe to answer "yes, and no." They face similar challenges and accomplish similar ends. It is reasonable to predict that, like the bacteriophages, they will exploit the full range of surface structures offered by their archaeal hosts. The most unique archaeal innovations for adsorption are likely to be found among those that infect extremophilic crenarchaea, paralleling Pharaoh's (STIV's) unique structures for virion egress (see "A Pyramid Builder" on page 182).

Two families[23] of filamentous crenarchaeal phage are known to carry specialized structures at both ends, structures that have been described variously as claw-like, mop-like, or resembling bottle brushes. These grappling hooks attach to the sides or tips of pili. Targeted cells can't escape infection by forgoing these appendages, as the phages then attach to the cell surface directly. One rod-shaped type[24] was found to quickly adsorb irreversibly to the tip of the pili-like structures of its host, then move down the side of the pilus toward the cell surface. There the virion disassembles (a trick shared by Skinny), and the DNA enters the cell. That's all that we know so far about how these phages enter their hosts. I'd love to know more.

Receptors Come and Receptors Go

Temperance's receptors, being part of E. coli's maltose uptake system, are abundant only when maltose is available in the environment, but at least a few are always present (see "Potential Receptors" on page 200). Both Nerd and Minimalist thrive despite using receptors that sometimes are completely absent. Nerd's Caulobacter host spends part of its life cycle as a sessile cell attached to the substrate by a stalk. The flagella, as well as Nerd's receptors nearby, are present only in the swimmer cells that bud off from the stationary mother cell. Minimalist's E. coli host engages in conjugation most often at 37° C, as when residing in a mammalian gut, and not at all below 25° C. The cells also forgo this form of sex when starvation halts growth. Thus, by using the conjugation pilus as its receptor, Minimalist automatically postpones infection until conditions improve and the cell can support phage replication.

Some phages routinely employ two RBPs, each of which uses a different receptor and gives the phage a different host range. One way to do that is to encode two different RBPs and include them in every virion. Alternatively, a phage could have the ability to synthesize either one

---

[23] the lipothrixviruses with their flexible filaments and the rudiviruses with their stiff rods.

[24] SIRV2, *Sulfolobus islandicus* rod-shaped virus 2.

of the two RBPs. This is the tactic of phage Mu.[25] Picture two genes for a key tail fiber protein side-by-side in the phage chromosome. Further picture that part of both genes lies within a segment of DNA that can be flipped, or inverted, within the chromosome.[26] The phage has a site-specific invertase that cuts the segment from the chromosome and pastes it back in, but in the opposite orientation. When the cassette is in one orientation, it directs the synthesis of an RBP that binds one receptor. Flip the cassette, and the RBP made now uses a different receptor, thus enabling the phage to infect different hosts. During each lytic cycle, most of the progeny phages inherit the parental orientation. Only rarely is the cassette inverted, but rarely is often enough to keep pace with shifting host availability.

The infection tactics of some bacterial pathogens, such as *Bordetella pertussis* (the causative agent of whooping cough), pose a recurring challenge to their phages. During its virulent phase, while invading someone's respiratory tract, *Bordetella* makes numerous virulence factors. These include pertactin, an OM protein that helps the bacterium to adhere to the epithelial cells lining our trachea. Pertactin is also the receptor of choice for Fickle (BPP-1).[27] While outside a host, *Bordetella* doesn't make pertactin, and thus would be expected to be immune to this phage. But it is not so easy to evade a phage. Fickle has a sophisticated system[28] that deals effectively with pertactin's periodic disappearance by occasionally mutating the RBP gene in a progeny genome. To do this, the phage carries two versions of the RBP gene. One version is the original pertactin-binding RBP that is archived as an immutable template. The other is the workaday copy used to synthesize the RBP.

---

[25] bacteriophage Mu, a myovirus, whose name refers to its ability to cause mutations in the host chromosome. This activity is unrelated to the gene inversion trick mentioned above. Mu inserts itself in random chromosomal locations, then later excises cleanly from that site and inserts somewhere else. Thus, it also qualifies as a transposable element, one of the groups of mobile genetic elements that will be discussed in PIC.

[26] invertible gene cassette: a chromosome segment whose orientation within the chromosome is periodically reversed. An invertase catalyzes the site-specific cutting of the segment from the chromosome followed by its pasting into the same location but with the reversed orientation.

[27] *Bordetella* phage BPP-1, a podovirus that infects various *Bordetella* species.

[28] diversity generating retroelement: a cluster of genes that introduces mutations in precise locations in a target gene, while maintaining the ability to restore the original sequence from an unaltered template copy. The gene cluster includes a reverse transcriptase that participates in the mutagenesis step.

Occasionally Fickle replaces the workaday copy with a new version made from the template and mutates the new rendition in the process. These mutations are restricted to the twelve amino acids that determine its host specificity. These twelve form the receptor-binding surface of the RBP in Fickle's globular "feet" (see Figure 80). Although not every mutation yields useful receptor-binding properties, one in a thousand to one in a million do. Some mutated RBPs enable Fickle to infect cells that lack pertactin. The

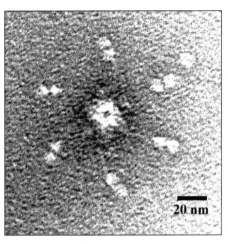

Figure 80: Fickle feet. TEM of the central hub of Fickle's short tail surrounded by six long tail fibers, each with a globular end. Fickle's mutable RBP is located in these globular structures. Courtesy of Mari Gingery, independent researcher.

phage can also restore its original pertactin-binding RBP by making a new copy from its template without any mutations. Each rendition persists for generations until a new version is made and replaces the old. This skillful targeted mutagenesis is not unique to Fickle and its RBP. It is employed by various phages, as well as by Bacteria, to modify other proteins where occasional changes in specific amino acids are a benefit. I think of this as yet another phage invention. Although that seems likely, we don't know for certain who thought it up first – the phages or their hosts.

The challenge of disappearing receptors can sometimes be avoided by selecting as your target an essential cell surface component, something that can't be jettisoned even temporarily, even when under intense phage attack. However, the prokaryote doesn't necessarily have to eliminate a receptor in order to make it invisible to the phage. Subtle modification, perhaps a single amino acid change in a protein of 200 amino acids, may be sufficient to cause a misalignment with the phage's RBP and prevent adsorption. As mentioned above, rarely does such tweaking come without a price. Bacteria, like phages, compete vigorously for their place in the biosphere. Every protein is the product of billions of years of evolution. Tinkering with some component to dodge a phage usually slows growth at least a little, under at least some conditions. Even an almost imperceptible effect, when amplified over the course of generations, makes resistant strains able to compete with their faster growing kin only when phage predation is intense. When this defense is successful, the population of infecting phage declines, which in turn allows the faster-growing, phage-sensitive strains to once again dominate the bacterial population.

Whatever the mechanism, resistance against phage attack is short lived. Soon a phage capable of using the modified receptor or equipped with some other work-around will evolve, and the "resistant" strain is prey once again. This seesawing between phage and host is perpetual, and its role in the maintenance of both phage and prokaryote diversity is major (see PIC).

From escape to adsorption, the virion's first mission is to keep its cargo safe and secure, and to bring it to the door of its new host. During that phase, success of any individual virion is a matter of chance. Will the virion collide with a potential host before it is irreparably damaged? Once it arrives at the door, expertise again becomes decisive. What had been an inert, passive transport vehicle springs into action. It adsorbs irreversibly to the cell and transforms itself into an active chromosome delivery device. The phage's slumber will soon end.

Further Reading

Baptista, C, MA Santos, C Sao-Jose. 2008. Phage SPP1 reversible adsorption to *Bacillus subtilis* cell wall teichoic acids accelerates virus recognition of membrane receptor YueB. J Bacteriol 190:4989-4996.

Barr, JJ, R Auro, M Furlan, et al. 2013. Bacteriophage adhering to mucus provide a non-host-derived immunity. Proc Natl Acad Sci USA 110:10771-10776.

Casjens, SR, IJ Molineux. 2012. Short noncontractile tail machines: Adsorption and DNA delivery by podoviruses. in *Viral Molecular Machines*: Springer. p. 143-179.

Davidson, AR, L Cardarelli, LG Pell, DR Radford, KL Maxwell. 2012. Long noncontractile tail machines of bacteriophages. in *Viral Molecular Machines*: Springer. p. 115-142.

Farley, MM, J Tu, DB Kearns, IJ Molineux, J Liu. 2017. Ultrastructural analysis of bacteriophage Φ29 during infection of *Bacillus subtilis*. J Struct Biol 197:163-171.

Garcia-Doval, C, MJ van Raaij. 2013. Bacteriophage receptor recognition and nucleic acid transfer. in *Structure and Physics of Viruses*: Springer. p. 489-518.

Guerrero-Ferreira, RC, PH Viollier, B Ely, JS Poindexter, M Georgieva, GJ Jensen, ER Wright. 2011. Alternative mechanism for bacteriophage adsorption to the motile bacterium *Caulobacter crescentus*. Proc Natl Acad Sci USA 108:9963-9968.

Howe, MM. 1980. The invertible G segment of phage Mu. Cell 21:605-606.

Hu, B, W Margolin, IJ Molineux, J Liu. 2013. The bacteriophage T7 virion undergoes extensive structural remodeling during infection. Science 339:576-579.

Hu, B, W Margolin, IJ Molineux, J Liu. 2015. Structural remodeling of bacteriophage T4 and host membranes during infection initiation. Proc Natl Acad Sci USA 112:4919-4928.

Kellenberger, E, E Stauffer, M Häner, A Lustig, D Karamata. 1996. Mechanism of the long tail-fiber deployment of bacteriophages T-even and its role in adsorption, infection and sedimentation. Biophys Chem 59:41-59.

Liu, M, R Deora, SR Doulatov, M Gingery, FA Eiserling, A Preston, DJ Maskell, RW Simons, PA Cotter, J Parkhill. 2002. Reverse transcriptase-mediated tropism switching in *Bordetella* bacteriophage. Science 295:2091-2094.

Medhekar, B, JF Miller. 2007. Diversity-generating retroelements. Curr Opin Microbiol 10:388-395.

Michel, A, O Clermont, E Denamur, O Tenaillon. 2010. Bacteriophage PhiX174's ecological niche and the flexibility of its *Escherichia coli* lipopolysaccharide receptor. Appl Environ Microbiol 76:7310-7313.

Romantschuk, M, VM Olkkonen, DH Bamford. 1988. The nucleocapsid of bacteriophage φ6 penetrates the host cytoplasmic membrane. EMBO J 7:1821-1829.

Samuel, AD, TP Pitta, WS Ryu, PN Danese, EC Leung, HC Berg. 1999. Flagellar determinants of bacterial sensitivity to χ-phage. Proc Natl Acad Sci USA 96:9863-9866.

Woody, M, D Cliver. 1997. Replication of coliphage Qβ as affected by host cell number, nutrition, competition from insusceptible cells and non-FRNA coliphages. J Appl Microbiol 82:431-440.

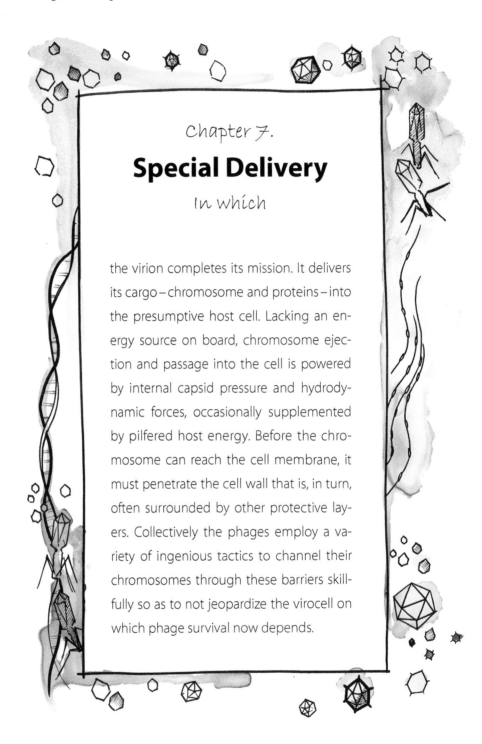

Chapter 7.

Special Delivery

In which

the virion completes its mission. It delivers its cargo – chromosome and proteins – into the presumptive host cell. Lacking an energy source on board, chromosome ejection and passage into the cell is powered by internal capsid pressure and hydrodynamic forces, occasionally supplemented by pilfered host energy. Before the chromosome can reach the cell membrane, it must penetrate the cell wall that is, in turn, often surrounded by other protective layers. Collectively the phages employ a variety of ingenious tactics to channel their chromosomes through these barriers skillfully so as to not jeopardize the virocell on which phage survival now depends.

You start where you can get an opportunity, you take everything that you can do to gain entrance.

Jon Voight

The entrance strategy is actually more important than the exit strategy.

Edward Lampert

A bacteriophage that ejects its DNA spuriously or into an unsuitable host cell ceases to exist.

C. Garcia-Doval and MJ van Raaij

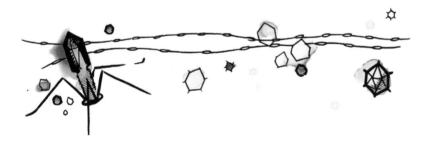

E arlier steps – the production of new phage chromosomes and structural proteins, virion assembly, and the quest for a new home – were all preparations for the grand finale: infection! Here the virion delivers its cargo, both chromosome and internal proteins, into a suitable cell. The care, precision, and quality control manifested in the preceding steps now pay off. For phages such as Lander (T4), chromosome delivery is flawless for 100 out of 100 virions.

Irreversible adsorption triggers the transformation of the inert chromosome shipping container into a chromosome delivery machine. The entire chromosome must be transferred, or the phage is dead. This is no small task. To picture the scale of this maneuver for a tailed phage such as Lander, imagine moving a few meters of tightly-coiled, cooked spaghetti through a narrow straw into a half-pint jar. Lander's ~52 µm long chromosome is transferred through the 4 nm wide channel inside its tail tube into an *E. coli* cell that is only ~1 µm long. Delivery speed also can matter. The sooner the chromosome gets to work, the sooner the progeny will be released. Lander is fast, delivering its 169 kbp chromosome in less than a minute, i.e., at a rate of 3,000–4,000 bp/s. This is remarkably fast. For comparison, when one bacterium transfers DNA to another through a conjugation pilus, the pace is a leisurely ~100 bp/s. Some phages, such as Stubby (see "Delivery with a Stubby Tail" on page 227), know that slower can be better. All tailed phages, as well as some without tails, use one specialized vertex for both DNA packaging and its subsequent delivery into the host. With only a very few exceptions (see "Shy's Delivery" on page 232), afterwards the empty virion is left abandoned outside.

Mobilizing Your DNA

What powers the movement of DNA out of the capsid and into the cell? Whereas DNA packaging necessitated major energy expenditure in the form of ATP consumed, delivery requires no cellular energy source.[1] Initially the source of the oomph was postulated to be the high pressure inside the capsid. According to this story, one need only

---

[1] As usual, there are a few exceptions that prove the rule. See "Delivery with a Stubby Tail" on page 227.

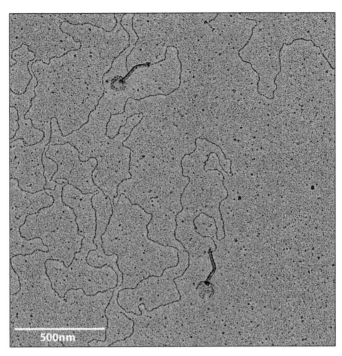

Figure 81: TEM of two Positivist virions caught in the act of chromosome release. Courtesy of Rudi Lurz Max-Planck, Institut for Molecular Genetics, Berlin, Germany, and Paulo Tavares, Institut de Biologie Intégrative de la Cellule, Gif-sur-Yvette, France.

poke a small hole in the capsid or pull the plug at the capsid portal, and the DNA would spew out of the capsid and into the cell. Phages with dsDNA chromosomes all package their DNA to about the same high density in their capsid (see "Packaging DNA" on page 124). Thus, they all have a pressure of tens of atmospheres available for this task. This model has been supported by *in vitro* studies with Temperance (λ) and others (see Figure 81). Temperance can be prompted to eject its chromosome *in vitro* by offering it liposomes[2] with its receptor embedded in the membrane. Under these circumstances, it ejects its entire 48.5 kbp chromosome into the interior of the vesicle in less than 1.5 seconds.

At first glance, this might seem to answer the question, but there are some hitches. The primary one: as the concentration of DNA in the

_____

[2] liposome: a small, membrane-bounded vesicle, typically artificially produced *in vitro*.

capsid decreases, so does the pressure propelling the DNA. The rate of ejection must decline as more and more of the DNA exits. This conflicts with experiments that demonstrated a constant rate of transfer. Further, the force of ejection is countered by the cell's turgor pressure (osmotic pressure), typically at least two atmospheres for *E. coli* and under some conditions more than ten. A cellular turgor pressure of two atmospheres is sufficient to halt DNA ejection when ~40% of the chromosome remains in the capsid. For Gram-positive *B. subtilis*, the turgor pressure is ~19 atmospheres. Internal capsid pressure could power the transfer of only half of Dynamo's (ϕ29's) DNA into its *B. subtilis* host. While this internal pressure can give the first portion of the chromosome a push, it can not do the whole job.

There are other issues with this pressure-driven model, as well. Only DNA chromosomes are packaged to such high densities, making a different mechanism necessary for phages whose chromosomes of RNA are more loosely packaged (see "Shy's Delivery" on page 232). It also clearly is not the whole story for some well-studied phages. Stubby (T7), for instance (see "Delivery with a Stubby Tail" on page 227), enters via a multistage process, including steps where the host actively pulls the chromosome inside.

A supplemental mechanism, applicable to almost all phages, is described by the "toilet flush" hydrodynamic model (see Figure 82). In an aquatic environment, osmotic pressure is continually trying to push water into the capsid due to the high concentration of DNA, proteins, and small solutes inside. Some water does enter, slightly expanding the capsid, but the capsid responds with an equal force to resist further influx and expansion. When the phage tail breaches the CM and the plug at the portal is removed, water streams from the environment into the capsid, down the tail tube, and into the cytoplasm. As water rushes into the cell, it brings the DNA – ultimately all the DNA – along with it by hydrodynamic drag. Picture a strand of cooked spaghetti going down the drain as you let the dishwater out of the sink. Since the amount of water entering the cell through this channel is too small to significantly alter the cell's osmotic concentration, the force driving the DNA remains constant and the entire chromosome is transferred at a steady rate. Any internal proteins packaged in the virion would

Figure 82: Hydrodynamic model. A schematic diagram showing a long-tailed virion before and after DNA ejection has begun. Note the flow of water through the capsid shell, down the tail tube, and into the cell's cytoplasm.

also be swept along with the DNA. This mechanism, described here for tailed phages, also can drive chromosome delivery for phages that don't package their chromosome so densely.

Barriers

Sweeping the DNA out of the capsid is necessary, but not sufficient, to allow the chromosome into the host-to-be. Whether Gram-positive or Gram-negative, Bacteria (and some Archaea) hide inside a wall of peptidoglycan. For a phage to infect these cells typically calls for a corridor through the peptidoglycan mesh that is large enough to allow passage of the phage's chromosome delivery device. A tailed virion may require only a corridor wide enough for its tail. The requirement is also modest for "tailless" virions that restructure to form a delivery tube during adsorption. Some others need a broad avenue wide enough for the entire virion to pass. Most phage virions are equipped with a peptidoglycan-degrading enzyme to clear away the nearby peptidoglycan. This dismantling of the wall must be kept highly localized. Too large a gap (i.e., more than ~20 nm for some Bacteria) and the CM, driven by turgor pressure, would bulge through the hole and eventually cause the cell to lyse. By digesting only the abutting cell wall, the virion deftly excavates a narrow tunnel as it moves toward the cell. Although each phage compromises merely a minimal region of the wall, lysis can result when many virions attack simultaneously, each one nibbling locally.

More tricky is the transfer of the chromosome across the CM. DNA and RNA are both long polymers that carry a strong negative charge along their entire length. The CM of all cells is a hydrophobic lipid bilayer that is impermeable to such molecules. What is needed is a hole in the membrane. A phage can't employ its holins (see "Scoring a Hole-in-One" on page 172) because they create large holes that would destroy the membrane potential or allow copious cell contents to leak out. What is needed for a dsDNA chromosome is a tidy ~2 nm diameter passageway. Many phages construct a precise membrane channel of that approximate size from either proteins brought along on board the virion or from host membrane components or from both. A few ions do leak out of the cell during chromosome delivery, but the loss is minimal, and immediately after transfer the hole is sealed.

Phages that infect Gram-negative Bacteria face four obstacles: the OM, the peptidoglycan, the CM, and the periplasm. They need not only channels through both membranes, but also a protective conduit more than 20 nm long to span the nuclease-infested periplasm in between.

Speculative Tails

The virions of some 5,500 different phages had been observed under the electron microscope by 2007. Approximately 95% of them were tailed phages (order *Caudovirales*). That this group is so successful, despite the cost of tail production, says that investing in a tail pays off. Are tails more effective or more efficient chromosome delivery mechanisms? Tailless phages often lack a specialized vertex for DNA transfer in/out. If the DNA can exit through any vertex, the phage can't predict which vertex will be used. As a result, it can't package its chromosome so that one particular end is always positioned at the door and is certain to enter the cell first. This matters a lot. Transcription of the genes that enter first can get underway long before the last gene is inside the cell. The first proteins synthesized can be protecting the DNA from host nuclease attack even before chromosome delivery is complete. In some instances, the mechanism of chromosome transfer itself requires specific sequences to initiate or regulate the process (see "Delivery with a Stubby Tail" on page 227). Tailless phages have short chromosomes that average 7,300 bp and rarely exceed 15,000 bp. By contrast, the average length among the tailed phages is 62,000 bp and the largest is nearly 500,000 bp.

Even among the long-tailed phages there is a bias regarding chromosome length. The chromosomes of the siphoviruses with non-contractile tails, such as Temperance, average 50,000 bp and few are more than 60,000. By contrast, the chromosomes of phages with a contractile tail average 110,000 bp. Getting such long chromosomes into and out of the virion flawlessly is challenging. Might the presence of a tail, and even its structure, influence or be influenced by chromosome length? Likely many factors are involved, but still I wonder.

Why are both contractile and non-contractile tails so very long? This question also remains unanswered. Granted, a virion docked at the outer surface of the cell envelope needs a conduit for DNA transfer that extends all the way to the CM, but these tails are far longer than required for that. Lander's contractile tail is 100 nm long and the non-contractile tails of the siphoviruses average 190 nm. By contrast, a Gram-negative cell envelope spans at most 40 nm, and even the thicker Gram-positive cell walls are only 80 nm. Conversely, are there any disadvantages to longer tails other than the increased cost of construction? They do present one challenge. When a virion adsorbs irreversibly to a potential host, interaction between the tip of the tail and the cell surface triggers DNA release through the capsid portal at the other end of the tail – a distance of 100 nm or more away. How is contact at the tip communicated up the tail to the capsid portal? Perhaps adsorption triggers a local conformational change in the tail proteins that propagates, domino-like, the length of the tail to release the DNA from the capsid. Alternatively, action at the tip may be sufficient. Often the tail tube is occupied by the tape measure proteins (TMPs) along with the end of the DNA chromosome that is destined to enter the cell first. Simply opening the tail tip might allow the TMPs to exit, with the

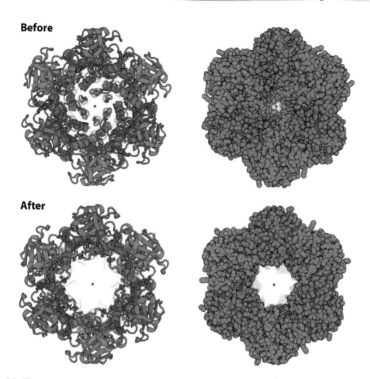

Figure 83: Dynamo's (φ29's) gate. Representations of a cross-section through Dynamo's tail tube near the tip both before and after DNA delivery. The six proteins that make up the tube are shown in alternating colors. Both a ribbon diagram and a space-fill drawing are provided for both cross-sections. The image before DNA delivery incorporates the structure of the tail knob protein as determined by X-ray crystallography at 3.5 Å resolution (Protein Data Bank ID 5FB5). A mutant protein (Protein Data Bank ID 5FEI, 2.6 Å resolution) in which that loop section had been deleted was used as a proxy for the structure after DNA delivery. Source: Protein Data Bank. Primary publication: Xu, J, M Gui, D Wang, Y Xiang. 2016. The bacteriophage φ29 tail possesses a pore-forming loop for cell membrane penetration. Nature 534: 544-547.

DNA following in their wake. We don't know yet which phages use these methods or which ones have devised yet other tricks.

Dynamo's Crafty Tail Knob

Since Dynamo's (φ29's) host is the Gram-positive *Bacillus subtilis*, it faces only two of the three barriers – the cell wall and the CM. For this phage, action at the tail tip may indeed be sufficient to prompt DNA release. Its tail is short, only 38 nm, and ends in a multi-functional knob that assembles from two different proteins. One protein[3] with

---
[3] gp13

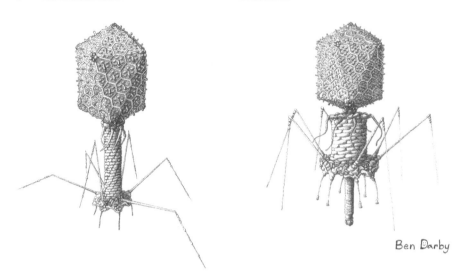

Ben Darby

Figure 84: Tail contraction. When a myovirus, such as Lander, delivers its chromosome, the proteins forming the outer tail sheath change their conformation and thereby cause the sheath to contract. The length of the inner tail tube remains unchanged. Since the LTFs and STFs anchor the baseplate to the OM, contraction pushes the tail tube through the OM. Original drawings of Lander by Ben Darby. Previously published in *Life in Our Phage World* by Rohwer, F, M Youle, H Maughan, N Hisakawa. 2014. Wholon. Used with permission.

peptidoglycan-digesting activity forms a protrusion at the very tip of the tail. When the 12 tailspikes have bound to their receptors, this enzyme digests a path for the virion through the 25 nm thick peptidoglycan cell wall. A hexamer of the other protein[4] forms the knob itself. Six of these elongated proteins are positioned in parallel to form a cylinder that surrounds a 4 nm wide channel. Although this opening is large enough to allow the dsDNA chromosome to exit, these same six proteins block the very exit that they have created (see Figure 83). Each has a loop that extends into the channel and folds back along two-thirds of the length of the rest of the protein. Having these six loops projecting into the channel reduces the clearance in the passageway to a mere ~0.6 nm, far less than required to accommodate DNA. During DNA ejection, these loops are in effect swept out of the channel and now form a hexameric, cone-shaped structure that extends the tail tube another 4 nm. This added length spans the CM where it creates a defined pore for DNA passage. Thus, with just these six protein molecules, Dynamo assembles a conduit for DNA passage, blocks that channel in

[4] gp9

the virion, opens the channel after adsorption, and extends the conduit through the CM to allow the DNA to enter the cell. Elegant efficiency!

Delivery by Lander

During delivery, Lander (T4) once again displays its expertise (see Figure 84). It consistently transfers not only its entire 169 kbp chromosome, but also the thousand internal proteins packaged along with the DNA inside its capsid. And it does this flawlessly and remarkably fast, in less than a minute. Delivery events begin when all six LTFs bind to the cell and bring the baseplate at the end of the tail into contact with the OM (see "Landing on the OM with a Long Tail" on page 198). This contact triggers a cascade of changes that starts with the conversion of the hexagonal baseplate into an expanded "star" configuration. Expansion opens the hole in the center that will later allow passage of the tail tube. Expansion also rotates the STFs so that they now point downward toward the OM. Binding of the STFs to the OM anchors the virion securely in position. The tail sheath, but not the tail tube inside, contracts. This contraction, combined with the syringe-like appearance of the virion, created the popular notion that a phage injects its DNA into the host like a spider injects venom into its prey. Vivid images like this persist, even though disproven.

Since the virion is firmly bound to the OM, sheath contraction pushes the tail tube through the OM (see Figure 85). This push does not require an external energy source. It is powered by the transition of all 144 proteins in the tail sheath to a more stable, lower energy conformation. The tail tube enters the periplasm where it encounters the peptidoglycan mesh of the cell wall. The tube is furnished with peptidoglycan-degrading activity that clears a narrow path and ultimately allows the tube tip to touch the CM. Sheath contraction does not push the tail tube through the CM. Instead, the tube induces the cell to assist. The CM bulges outward and fuses with the tip of the tube to form a transmembrane channel for DNA entry. The net result: the tail tube forms a continuous protective conduit for the chromosome from capsid to cytoplasm. First out of the virion and into the cell are the TMPs, followed by the other internal proteins and the DNA. Lander has arrived, with its protein entourage, and an infection is underway.

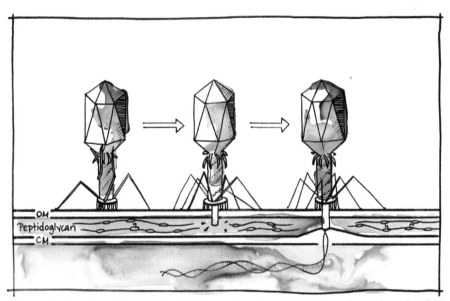

Figure 85: Lander delivers. (Left) Two of Lander's LTFs have bound to the OM and the STFs have rotated to point downward. (Center) With the virion now firmly anchored to the cell wall by all the LTFs and STFs, the sheath contracts and pushes the tail tube through the OM. (Right) The CM bulges outward to meet and fuse with the tail tube, and the DNA exits into the cytoplasm.

Delivery with a Stubby Tail

Lander's method works well, but not every phage infecting a Gram-negative host is a myovirus with a contractile sheath, nor do they all transfer their chromosome so quickly. Nevertheless, they all get the job done. They all safely deliver their chromosomes across the periplasm and through the CM. Podovirus Stubby (T7) does not have a contractile tail sheath to facilitate the process, and the short tail (23 nm) on its virion isn't quite long enough to reach from the OM to the CM. Actually, Stubby's completed tail is long enough, but Stubby doesn't assemble its full length until needed (see Figure 86). It packages additional tail components inside the capsid adjacent to the exit portal and assembles them outside the capsid to extend the tail to 40-55 nm after adsorption.

Closest to the portal and the first to exit are ten molecules of the protein[5] that forms the channel through the OM. Following them out of

---

[5] gp14

Figure 86: Stubby delivers. After adsorption Stubby assembles the core proteins carried inside the virion to extend its tail and form a conduit from capsid to cytoplasm.

the capsid are two different proteins,[6] eight of one and four of the other, that are released into the periplasm. One attacks the peptidoglycan to clear a path to the CM. The two assemble together to form the tail extension that spans the periplasm and penetrates the cell membrane during chromosome delivery. These same two proteins use cellular energy to ratchet in the first 2% of the chromosome at a leisurely 70 bp/second. This part of the chromosome is transcribed immediately by the cell's RNAP. Early gene transcription is a key part of Stubby's strategy for survival on arrival because one of these first genes encodes its RE (restriction endonuclease) inhibitor (see "Stubby's Endonuclease Inhibitor" on page 72).

The remainder of the chromosome is pulled into the cell by transcription. Normally, during transcription the RNAP moves along a comparatively stationary DNA strand. When RNAP proceeds in that manner here, it runs into a wall when it meets the CM. There is more DNA to be transcribed, but that DNA lies on the other side of the membrane. Although RNAP is blocked from advancing further along the chromosome, it nevertheless continues to transcribe one base pair after another by pulling the DNA into the cell as it works. The next 18% of the chromosome is brought in this way by the host's RNAP at a sluggish 40–50

[6] gp15 and gp16

bp/second. One of the genes transcribed in this 18% encodes Stubby's own faster RNAP. Its transcript is promptly translated and the new phage RNAP pulls in the remaining 80% of the chromosome at 250 bp/second. Even that faster rate is ten times slower than the delivery rate for Lander's chromosome. Why go so slow? Here's one reason. Lander packages its RE inhibitor inside the capsid for delivery with the chromosome. Thus, its inhibitor is on duty immediately. Stubby instead relies on slower delivery to allow time for synthesis of its RE inhibitor before most of the chromosome is at risk in the hostile cytoplasm.

Stubby's way is not a one-size-fits-all method readily employed by any phage. Specific sequences are required in the chromosome to initiate and terminate transcription by the host RNAP, and then to start transcription by Stubby's own RNAP to finish the job. In order for those sequences to be at the correct position in every chromosome, the phage must package identical chromosomes in every virion and must also ensure that the same chromosome end exits first every time. Stubby meets these requirements, while Lander and many others do not. Moreover, transcription of all of Stubby's genes gets underway during delivery, whereas many phages don't transcribe their middle and late genes until long after the entire chromosome is inside the cell. As noted above, Stubby is one phage that requires cellular energy to power DNA movement. However, this tactic does not cost Stubby extra cellular energy because every phage transcribes its chromosome during infection. Stubby's chromosome delivery simply piggy-backs on necessary phage gene transcription.

Yoda's Grand Entry

It is possible for a phage to deliver a chromosome across two membranes and the periplasm without a virion tail, either short or long. For proof, consider Yoda's (φX174's) tactics. Its virion is a simple icosahedral capsid with a prominent mushroom-shaped spike at each vertex. Yoda tethers its packaged chromosome, a small circle of ssDNA, to each of the 60 capsid proteins. Since the chromosome is circular, there is no free end to be parked at one particular vertex ready to exit. Apparently, there is no dedicated exit portal. All of the 12 spikes may be capable of adsorption and chromosome delivery. Each has a central channel that extends its entire length and is spacious enough to allow Yoda's chromosome to exit.

Without a tail, how does Yoda send its chromosome safely into its *E. coli* host? For this task, and much more, it relies on its multi-functional "pilot" protein. This is the protein that recognizes and adsorbs to potential hosts, and later at least one copy escorts the chromosome into the cell. Twelve copies are present in each virion. Although one is associated with each spike during assembly, their location in the mature virion remains uncertain. After adsorption, with the virion moored to the OM, ten of the pilot proteins form a tube that is extruded from the docked vertex. The ten are aligned in parallel with one of their ends anchored in the OM, the other in the CM. In between, they form a coiled coil structure that bridges the periplasm and contains a central passageway large enough to accommodate two strands of ssDNA with intercalated bases, i.e., with each base in one strand inserted between two bases in the parallel strand. Why two strands from Yoda's single-stranded DNA chromosome? Its chromosome is circular. Picture a rubber band moving through a straw. In this manner, Yoda's simple tunnel, assembled from a mere ten copies of a single protein, conveys its chromosome across the periplasm and into the cell seemingly as effectively as Lander's sophisticated tail tube.

Lipid Opportunities: Membrane Fusion

Those phages – clearly only a small minority – that incorporate a lipid membrane in their virion have available an option that is nonexistent for the majority. Given the right circumstances and the assistance of specialized membrane proteins, two lipid membranes can fuse. Those two membranes can be a virion membrane and the CM. Topologically, fusion of two membrane-bounded cells produces a single cell. If one cell is much smaller than the other, you could think about this as the smaller entering the larger. If a virion wrapped in a lipid membrane fuses with a CM, that virion is now inside the cell. This is a common tactic among animal viruses, but is not widespread in the phage world for at least two reasons. First, prokaryotes with their cell walls and other envelope components are not as accommodating as are membrane-bounded animal cells that routinely take in virion-sized particles by endocytosis.[7] Second, adding a membrane to a phage capsid complicates its construction and precludes adding a tail during assembly (see "A Lipid Supplement" on page 144). Nevertheless, a few phages exploit this option to make their way into their bacterial hosts.

One such phage is Fusion (PM2). Its virion appears from the outside to be a rather ordinary icosahedral protein shell ~65 nm in diameter with spikes at each vertex (see "Figure 56" on page 144). However, having such a large capsid for a 10 kbp dsDNA chromosome signals that there is something unusual about this virion. Indeed, Fusion's highly supercoiled, circular chromosome is loosely packaged inside, so loosely that it exerts no pressure on the surrounding protein shell – one more bit of proof that not all phages use the pressure within their capsid to transfer even part of their chromosome. In addition, the DNA is contained within a membrane sac that carries out important tasks during chromosome delivery.

An infection begins when a spike on Fusion's capsid recognizes and binds its receptor on the OM of its Gram-negative host (*Pseudoalteromonas*). The chromosome within is not ejected into the cell, but rather the protein capsid falls apart to expose the lipid sac. Quickly the phage membrane fuses with the OM, thereby becoming part of the OM and depositing the naked DNA into the periplasm. Here the chromosome is exposed to the resident nucleases, but resists destruction. Being a circular molecule of dsDNA, it evades attack by the exonucleases, and its supercoiled structure may provide additional protection from some endonucleases. Perhaps defensive proteins were also included inside the virion along with the chromosome. The region of the OM derived from the phage membrane now faces the peptidoglycan. One of the eight different phage proteins embedded in Fusion's membrane likely digests the peptidoglycan locally, thereby helping the DNA to access the CM. To cross that membrane the naked DNA may hijack the DNA uptake mechanism of its host. Why does Fusion's *Pseudoalteromonas* host, as well as some other Bacteria, have a mechanism to actively im-

---

[7] endocytosis: an active process carried out by all eukaryotic cells to acquire molecules or small particles that cannot penetrate the CM. A small region of the CM invaginates and then pinches shut at the neck to form a vesicle in the cytoplasm that contains a small volume of the extracellular fluid.

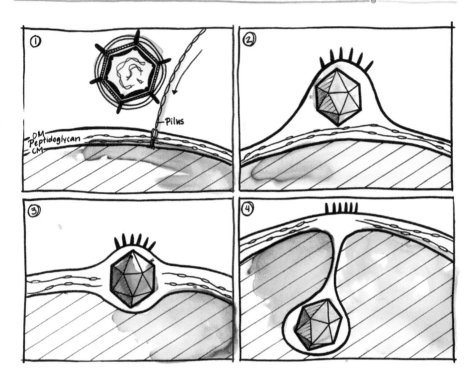

Figure 87: Shy Enters. When Shy (φ6) has adsorbed to the sides of a host pilus with its spikes, pilus retraction brings the virion to the OM. There its membrane envelope fuses with the OM, which dumps the capsid into the periplasm. Shy digests a path through the peptidoglycan to come face-to-face with the CM. There it induces the CM to invaginate and bud off a vesicle that carries the enclosed capsid into the cytoplasm.

port foreign DNA into its cytoplasm? DNA released by ongoing bacterial cell death is abundant in aquatic environments, even more so in biofilms and sediments. Some Bacteria recognize and actively acquire DNA from conspecifics that provides a source of potentially useful genes. Being a rich source of phosphorus and nitrogen, any available DNA makes a nutritious snack for any bacterium. Cellular DNA uptake systems offer an opportunity that you can be sure some phage, perhaps Fusion, exploits for a free ride inside.

Like Fusion, Shy (φ6) relies on membrane fusion to cross the OM of a Gram-negative bacterium (*Pseudomonas syringae*), but every other step in its delivery process is distinctly different (see Figure 87). Recall that this phage has three chromosomes composed of dsRNA (see "The Plight of RNA Chromosomes" on page 83). Cells regard molecules of dsRNA, when they are gene-length or longer, as undeniably foreign and destroy them promptly. To avoid that fate, Shy delivers its chromosomes still packaged inside a protein shell and never exposes them, unprotected, to the cytoplasm. When outside a host, these chromosomes travel inside a complex capsid composed of an inner and an outer protein shell, all of which is enveloped by a membrane. Protein spikes that project through the membrane effect host recognition and adsorption. Shy's host is a plant pathogen that uses specialized pili to deliver toxic molecules into plant cells. Shy, in turn, uses that same pilus to infect its pathogenic host. Its aphorism: those who infect by the pilus shall perish by the pilus. Like other pili, these structures go through alternating periods of extension and retraction by the addition and removal of their protein subunits at the cell membrane. When one of Shy's virions has attached to the side of a pilus by its spikes, pilus retraction brings it through the cell's protective polysaccharide capsule directly to the OM. The virion membrane fuses with the OM, thus bringing the double-layered capsid into the periplasm. Next barrier: the peptidoglycan cell wall. A peptidoglycan-degrading enzyme in the outer capsid clears a path for the virion through the peptidoglycan mesh to the CM. Virion contact with the CM induces a response that is very unusual for a bacterial membrane. Similar to endocytosis by eukaryotic cells, the membrane invaginates and forms a deep pit with the virion inside. The indentation constricts at the neck and pinches off the bottom of the pit to form a vesicle. This brings the virion "into" the cell, but now it is trapped inside a membrane-bounded vesicle. How it escapes is not known, but escape it does, shedding its outer capsid in the process. At this point the chromosomes, inside the protective inner capsid, can get to work.

Lipid Opportunities: Conduit Construction

In the previous examples, the phages used proteins to construct a conduit to facilitate entry of their chromosome into a cell. These protein conduits included Lander's complex tail built during virion assembly and Yoda's simple tube extruded after adsorption. There is an alternative construction material available: lipid membranes with a high protein content. The benefits of adding a lipid layer are not limited to fusion with a host membrane or to enhanced capsid strength and resilience.

Figure 88: Central tomographic section of a Slick (PRD1) virion with its protruding chromosome delivery chute. Red arrows mark the portal; the black arrow points to the unopened conical tip of the tube; the purple arc indicates the location of the chromosome that has not yet entered the tube. Reproduced with permission from Peralta B, Gil-Carton D, Castaño-Díez D, Bertin A, Boulogne C, et al. (2013) Mechanism of Membranous Tunnelling Nanotube Formation in Viral Genome Delivery. PLoS Biol 11(9): e1001667. doi:10.1371/journal.pbio.1001667.

Lipid-based structures can also undergo dramatic, rapid transformation and actively assist with delivery, as Slick (PRD1) demonstrates.

Like Yoda, Slick assembles a tailless icosahedral virion and forms its delivery tube post-adsorption from components carried inside the virion. However, its capsid is far more complex than Yoda's. To wit, it contains 18 different structural proteins, an internal lipid membrane rich in embedded proteins, and a specialized portal vertex dedicated to chromosome packaging and delivery. DNA is packaged to high density inside the membrane sac, with the membrane itself stretched and tacked to the interior of the capsid shell (see "Slick Assembly" on page 148).

The receptor binding proteins (RBPs) are on the spikes. Upon contact with a host, one spike recognizes Slick's receptor in the *Salmonella* OM and then binds to it weakly and reversibly. Picture Slick's icosahedral virion rolling along on the OM, one vertex spike after another contacting the cell surface. Each adheres only long enough and firmly enough to keep the virion from drifting away from the surface. The twelfth vertex, the portal vertex, is the site of irreversible adsorption and delivery. When a favorable roll aligns the portal vertex with a receptor, the virion adsorbs irreversibly and that vertex responds. Adsorption there decaps the neighboring vertices by removing the proteins that were blocking the 6 nm holes at each vertex. This opens the interior of the capsid to the external aqueous environment. Decapping also severs the connections between membrane and capsid at these points. The changing internal osmotic concentration induces major conformational changes to at least some of the membrane proteins. Driven by these changes, the membrane transforms from an icosahedral sack into a tube that initially protrudes from the portal vertex and then grows into a 50 nm long chute (see Figure 88). The tube is closed at its tip as it passes through the OM and digests a path through the peptidoglycan. When the tip contacts the CM, it opens and the chromosome passes into the cytoplasm.

The lipid membrane has demonstrated its usefulness by providing a DNA conduit across the periplasm and through both membranes, but it does more. It also generates some of the force required to deliver the entire chromosome into a cell against the resistance offered by the cell's turgor pressure. The phage membrane lining the capsid reduces the space available for the DNA, thus increases the pressure resulting from compaction of the defiant DNA. This internal pressure will be harnessed to drive the initial phase of chromosome delivery. In addition, the membrane was stretched by packaging of the DNA inside, like an inflated balloon. When the tube tip opens and DNA can exit the capsid, the membrane exerts an additional push as it deflates. As the DNA exits the capsid, further shrinking of the sac apparently maintains some force on the remaining DNA to ensure delivery of the entire chromosome. This shrinkage might be due to ongoing restructuring of the membrane or loss of membrane material through the tube tip. Although this lipid-assisted delivery mechanism performs superbly, it has been found so far in only Slick and a few of its very close kin.

Lipid Opportunities: Metamorphosis

A phage can work lipid magic even if it does not have a discernible membrane in its virion. Spindly (His1)[8] covalently links lipid components to its capsid proteins, thereby converting them into proteolipids. This gives its spindle- or lemon-shaped capsid more structural flexibility. Some of Spindly's lemons are somewhat shorter than others, but the shorter ones are also wider – more than 40 nm wide versus less than 35 nm. The net result: they all, shorter or longer, provide about the same cargo space for the chromosome. After adsorption, these capsid proteolipids perform more magic. As the DNA exits, each lemon – at most ~92 nm long and ~40 nm wide – morphs into a uniform tube ~138 nm long and only ~22 nm wide (see Figure 89). The diameter

---

[8] Spindly: halo archaeal virus His1, an unclassified phage with a spindle-shaped virion.

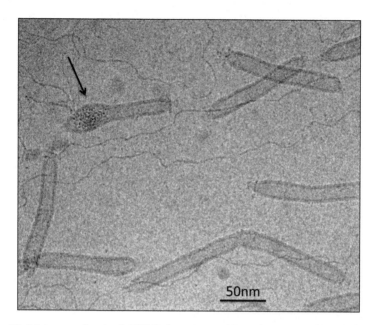

Figure 89: Metamorphosis. A TEM of a group of Spindly (His1) virions after they have ejected their chromosome, transforming from a spindle to a tube in the process. The arrow highlights one lemon caught in the act. The released DNA is visible as long thin filaments in the background. Courtesy of Chuan Hong and Wah Chiu, both at Baylor College of Medicine.

of the central channel is more than ample to accommodate transit of the dsDNA chromosome. As is the case for so many intriguing phages that infect Archaea, details of the story—such as how the chromosome passes from tube to cytoplasm—await investigation.

Theft of Intellectual Property

Phage tails are such sophisticated, deadly delivery devices that it isn't surprising that some of their bacterial targets have co-opted them for their own purposes. Starting with basic tail components, these Bacteria have fashioned anti-bacterial weapons known as tailocins. It is easy for Bacteria to acquire the cluster of phage genes that encodes the tail structural proteins of a temperate phage. During lysogeny, these phages insert their entire chromosome into the host chromosome as a prophage (see "Prophage Integration" on page 245). While residing there, the prophage is replicated along with the rest of the chromosome and thus is subject to the same low rate of mutation from replication errors. When such a mutation renders a prophage incapable of excising,

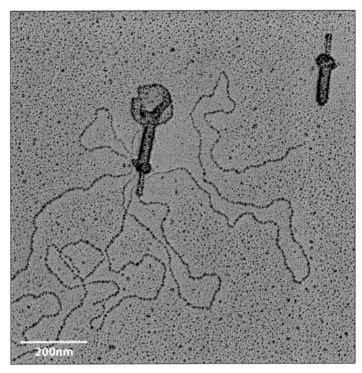

Figure 90: Delivery by a contractile tail. Some tailocins are contractile tails that were co-opted and adapted by Bacteria to deliver toxins or other cargo. Their original function for the phage was chromosome delivery, as shown in this TEM. Here, phage A511 was caught in the act of releasing its 134,494 bp dsDNA chromosome through its contractile tail. Upper right: a contracted tail broken from the capsid during sample preparation. Courtesy of Rudi Lurz, Max Planck Institute for Molecular Genetics, Berlin, Germany.

the prophage is marooned as an island in the host chromosome even though most of its genes remain intact. Gradually the cell eliminates those useless genes from its chromosome. However, various bacterial species found a use for a few of those "useless" genes. They retained the tail gene cluster and subsequently modified it to serve them as a tailocin. Like the ancestral tails, released tailocins have the ability to recognize and adsorb to specific Bacteria. Some were also modified to carry and deliver a small toxic payload. A single tailocin can kill a bacterium by either delivering a toxin or by puncturing the CM.

Both contractile (see Figure 90) and non-contractile phage tails have been converted into tailocins by Gram-negative Bacteria. Most strains of *Pseudomonas aeruginosa*, for example, make tailocins of one structural type or the other, and some have one or more of both types. Under the electron microscope, both types look like typical phage tails, complete with tail fibers. As with phage tails, selective adsorption to specific Bacteria is mediated by the RBPs on the tail fibers. For the contractile type, tailocin adsorption is followed by sheath contraction that forces the tail tube through the OM and then punctures the CM. Death follows quickly due to loss of the electrical gradient across the membrane. For self-protection, tailocin producers evolve immunity to their own particular tailocin so that they are not potential targets for the tailocins produced by their siblings.

The tailocin producers also retained the phage genes for lysis and use them to release assembled tailocins. Thus, a bacterium that deploys its tailocins is a dead bacterium. Tailocin production is reserved for use only as a last resort when the cell is faced with irreparable DNA damage. Given that deployment is suicide, what is the benefit to the producer? This is not clear. Some tailocins kill conspecifics that lack the same tailocin, some target other bacterial species, and others bring down big game many times the producer's size. For example, in the pastures of New Zealand, the New Zealand grass grubs[9] spend the summer months several inches below the soil surface munching on the roots of grasses and other crops. Although now a bane to farmers, the larvae have long provided a nutrient-rich habitat for the enterobacterium *Serratia entomophila*. Infection of the larvae by pathogenic strains of *S. entomophila* causes the larvae to stop feeding within 48 hours and to eventually die of starvation. We don't know the details of this pathogen's attack strategy. One likely scenario: when these pathogenic Bacteria arrive in the larval gut, tailocin production is switched on in a subset of the invaders. These tailocin producers release tailocins that enable their non-producing kin to colonize the insect gut, sacrificing themselves in the process. However they do it, these pathogens are a natural ally for modern day farmers in their battle against the grubs. Farmers have extended the chain of exploitation and now use *S. entomophila*, armed with its tailocins, as a biocontrol agent.

How did these contractile tailocins, related to Lander's tail and originally adapted to target bacterial cells, come to now attack eukaryotic

---

[9] the larvae of a scarab beetle (*Costelytra zealandica*)

insect cells? In an ancestor of today's *S. entomophila,* the original RBP in the tailocin tail fibers was replaced by one from a virus that infects eukaryotes (avian adenovirus). This type of gene swapping is highly plausible. Similar exchanges of RBPs between phages are evident in phage genomes where they enabled these phage to evolve a new host range (see PIC). For *S. entomophila,* this gene swapping was one of the critical steps that enabled these Bacteria to convert insect larvae up to 20 mm (~3/4 inch) long into a richly-endowed culture vessel.

These delivery mechanisms are admirably creative and adapted to exploit the materials provided by both the virion and the recipient cell. Nevertheless, none guarantee success. Every phage chromosome is delivered into hostile surroundings. Host defenses stand ready to cleave the chromosome into fragments or to interfere with phage replication in other ways (see "Survival on Arrival" on page 61). Many phages perish at every stage of the journey, but enough survive to repeat the cycle once again, as they have done for billions of years.

But wait! This is not the only story. Some phages eschew this never ending treadmill, this wasteful destruction of the virocell, this cycle of rapid replication leading immediately to homelessness and the quest for the next host. They pursue another option in which phage and host cohabitate, even collaborate for the common good. Eventually, the phage opts out and sees to its own replication. Who wins here? Who loses? Read on.

Further Reading

Hong, C, MK Pietilä, CJ Fu, MF Schmid, DH Bamford, W Chiu. 2015. Lemon-shaped halo archaeal virus His1 with uniform tail but variable capsid structure. Proc Natl Acad Sci USA 112:2449-2454.

Hu, B, W Margolin, IJ Molineux, J Liu. 2013. The bacteriophage T7 virion undergoes extensive structural remodeling during infection. Science 339:576-579.

Hu, B, W Margolin, IJ Molineux, J Liu. 2015. Structural remodeling of bacteriophage T4 and host membranes during infection initiation. Proc Natl Acad Sci USA 112:4919-4928.

Hurst, MR, SS Beard, TA Jackson, SM Jones. 2007. Isolation and characterization of the *Serratia entomophila* antifeeding prophage. FEMS Microbiol Lett 270:42-48.

Kivelä, HM, R Daugelavičius, RH Hankkio, JK Bamford, DH Bamford. 2004. Penetration of membrane-containing double-stranded-DNA bacteriophage PM2 into *Pseudoalteromonas* hosts. J Bacteriol 186:5342-5354.

Michel-Briand, Y, C Baysse. 2002. The pyocins of *Pseudomonas aeruginosa*. Biochimie 84:499-510.

Molineux, IJ, D Panja. 2013. Popping the cork: Mechanisms of phage genome ejection. Nat Rev Microbiol 11:194-204.

Peralta, B, D Gil-Carton, D Castaño-Díez, A Bertin, C Boulogne, HM Oksanen, DH Bamford, NG Abrescia. 2013. Mechanism of membranous tunnelling nanotube formation in viral genome delivery. PLoS Biol 11:e1001667.

Poranen, MM, R Daugelavicius, DH Bamford. 2002. Common principles in viral entry. Ann Rev Microbiol 56:521-538.

Roos, W, I Ivanovska, A Evilevitch, G Wuite. 2007. Viral capsids: Mechanical characteristics, genome packaging and delivery mechanisms. Cell Mol Life Sci 64:1484-1497.

Santos-Pérez, I, HM Oksanen, DH Bamford, FM Goñi, D Reguera, NG Abrescia. 2016. Membrane-assisted viral DNA ejection. Biochim Biophys Acta 1861:664-672.

Sun, L, LN Young, X Zhang, SP Boudko, A Fokine, E Zbornik, AP Roznowski, IJ Molineux, MG Rossmann, BA Fane. 2014. Icosahedral bacteriophage φX174 forms a tail for DNA transport during infection. Nature 505:432-435.

Xu, J, M Gui, D Wang, Y Xiang. 2016. The bacteriophage φ29 tail possesses a pore-forming loop for cell membrane penetration. Nature 534:544-547.

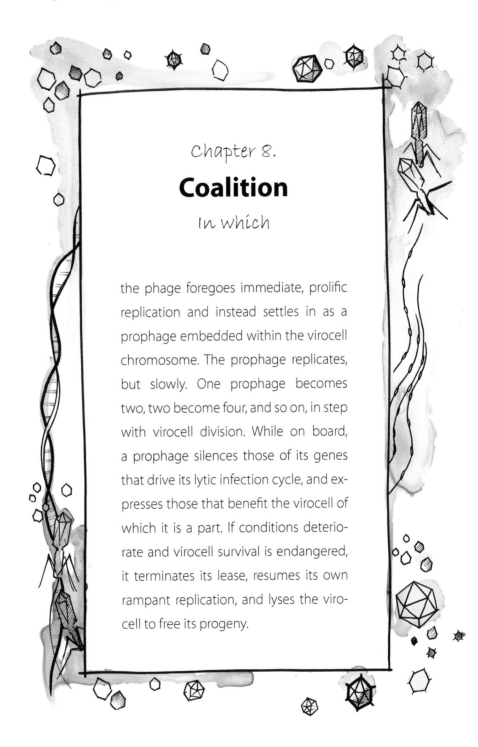

Chapter 8.

Coalition

In which

the phage foregoes immediate, prolific replication and instead settles in as a prophage embedded within the virocell chromosome. The prophage replicates, but slowly. One prophage becomes two, two become four, and so on, in step with virocell division. While on board, a prophage silences those of its genes that drive its lytic infection cycle, and expresses those that benefit the virocell of which it is a part. If conditions deteriorate and virocell survival is endangered, it terminates its lease, resumes its own rampant replication, and lyses the virocell to free its progeny.

Lysogeny is the hereditary power to produce
bacteriophage. A lysogenic bacterium is a
bacterium possessing and transmitting the
power to produce bacteriophage…Prophage
is the form in which lysogenic bacteria
perpetuate the power to produce phage.
André Lwoff 1953

Thus, prophages are not solely dangerous
molecular time bombs that can bring about
cell mortality, but also serve as a key to
bacterial survival in the oligotrophic oceans.
John Paul 2008

Politeness is the poison of collaboration.
Edwin Land

Any coalition, especially where one party
is more powerful than the other, it's always
bound to have a pecking order.
Peter Hook

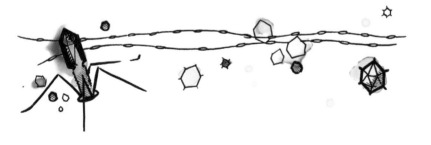

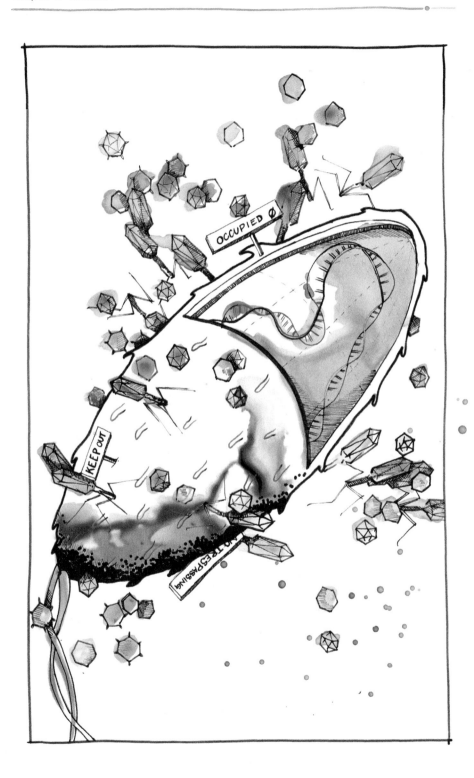

A re takeover and immediate replication always the best infection tactics for a phage? Rapid production and release of abundant progeny virions is one way to counter the low probability of a successful quest, but might there be a better way to utilize host resources, at least under some conditions? Many phages, possibly the majority, say "yes." These are the temperate phages. Each time a temperate phage arrives in a host cell, it chooses between the two alternative pathways available to it: lysis and lysogeny. Like the strictly lytic phages, it can immediately launch a lytic infection complete with the usual chromosome replication, production of phage proteins, assembly, and ultimate virocell destruction by lysis. On the other hand, it can opt for the lysogenic alternative – a substitute for the single-minded pursuit of maximum short-term gain. Here the phage resides quietly inside the virocell for an indefinite period of time while the newly formed phage-host coalition perks along. The virocell, now called a lysogen,[1] continues to grow and can give rise to multiple generations of descendants, each with a phage on board. Meanwhile, this resident phage, termed a prophage, considers that it has committed to only a moment-to-moment lease. When conditions threaten the virocell's viability, the prophage opts out and resumes lytic replication.

Note: This chapter focuses almost exclusively on temperate phages that infect Bacteria. Although many archaeal phages are non-lytic, we know little about their cooperative ways.

Unmasking Lysogeny

Nearly a hundred years ago, early phage researchers observed something very puzzling. Some bacterial strains cultured in the lab, apparently virus-free, could give rise to phages. Such strains were termed lysogenic, i.e., capable of lysing and releasing phage. What was going on? Clues accumulated. The researchers could convert a non-lysogenic strain into a lysogenic one by infecting a culture with one of these phages. However, when they ruptured the lysogens and looked inside for phage, they found no infectious particles. Their conclusion? The phage

[1] lysogen: a virocell with one or more resident prophages.

persisted in the cells in some non-infectious form until something prompted them to produce new particles. More evidence of their intracellular persistence: the phage particles released by these lysogens were new entities, but they were always of the same type as those that had originally infected that bacterial culture generations ago. Apparently whatever produced those virus particles had become part of the cell's heredity. Also, the lysogens could be routinely induced to release infectious particles on demand by damaging their chromosomes with UV irradiation or chemical mutagens. This was all quite mysterious. Bear in mind that this initial sleuthing was done before DNA had been identified as the genetic material in cells.[2]

Participation in a prophage–lysogen coalition provides benefits for both parties (see "Why Be Temperate?" on page 260 and "What's in It for the Bacterium?" on page 261). The lysogen shelters the prophage from environmental hazards and replicates it as part of its own chromosome. Each time the cell divides, both daughter cells inherit a chromosome with a hitchhiking prophage. Thus, one prophage becomes two, two become four, and so on, at a modest, steady pace while avoiding the hazards of the quest for a host. As rent, the prophage not only silences its genes that would launch lytic replication, but it expresses others that increase virocell fitness. Benefiting the host is not altruism. The prophage's own survival depends on the survival and well-being of its virocell.

Prophage Integration

The typical temperate phage inserts its dsDNA chromosome into the cell's chromosome where it resides as a chromosomal prophage.[3] In-

---

[2] For two key papers on the role of DNA as the hereditary material, see Avery et al. 1944 and also Hershey and Chase 1952 in "Further Reading" on page 59.

[3] Phages with ssDNA chromosomes aren't excluded from lysogeny. After arrival, their ssDNA is converted to dsDNA that can then integrate into the host chromosome using either their own integrase or host enzymes. To the best of my knowledge, no temperate phages with RNA chromosomes are currently known. However, eukaryotic ssRNA viruses, such as HIV, do insert into cellular chromosomes after they have converted their RNA chromosome to DNA via reverse transcription. Since some phages encode reverse transcriptase, the discovery of a temperate RNA phage would delight, but not astound, me.

serting at random locations would risk disrupting a gene, thereby causing a deleterious, heritable mutation. Prophages that inserted so indiscriminately were winnowed out by natural selection.[4] Each prophage inserts at a specific location in the host's chromosome where it remains until it decides to go lytic. Frequently that site is within a tRNA gene, but sometimes it lies within an intergenic region or even within a gene.

The consistent selection of the prophage's insertion site is achieved by site-specific recombination[5] between the host's circular dsDNA chromosome and the circularized phage dsDNA chromosome (see Figure 91). Both the phage chromosome and the bacterial chromosome possess specific "attachment" sites (ATTP and ATTB, respectively) that are recognized by that particular phage's integrase[6] enzyme. The presence of these corresponding sites and a phage integrase that recognizes them attests to the long evolutionary history of this intimate coalition. The integrase brings the ATTP and ATTB sites close together, then makes staggered cuts in both, and lastly joins the phage chromosome ends to the free ends of the bacterial chromosome. *Voilà!* The prophage now constitutes an island of typically about 50,000 bp within a chromosome of several million. Each time the host's DNA polymerase replicates the host chromosome, the prophage island is included. By contrast, when host genes are transcribed by the host's RNAP, only some of the prophage genes are expressed. Most often it is the prophage that regulates transcription of its own genes. During lysogeny the prophage expresses its repressor protein that silences the genes responsible for the lytic pathway, while also expressing others that thwart infection by related

---

[4] Some do get away with this disregard – for example phage Mu. Some other mobile genetic elements do likewise, specifically some of the transposons. Transposons are a diverse family of mobile genetic elements that can move from one site to another within or between chromosomes. They are typically short DNA segments that encode only a few proteins in addition to those required for transposition. Some transposons integrate at specific sites, while others can use any location.
[5] site-specific recombination: a form of recombination that occurs between specific chromosomal sites. It requires a short region of sequence homology between the participating chromosomes. A specific enzyme, such as the phage integrase, catalyzes the cleavage of both chromosomes, exchanges the strands, and then rejoins the ends.
[6] integrase: the enzyme encoded by temperate phages that catalyzes prophage insertion into a host chromosome by site-specific recombination. This integrase is distinct from the retroviral integrase that acts on DNA produced by reverse transcription of an RNA virus and inserts it into the DNA of eukaryotic cells.

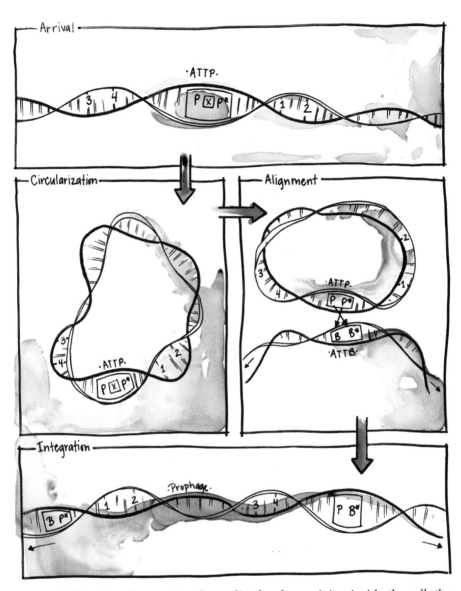

Figure 91: Prophage integration. Immediately after arriving inside the cell, the linear chromosome of a temperate phage circularizes via the complementary single-strand overhangs at each end. To integrate, its ATTP sequence aligns with the corresponding sequence in the bacterial chromosome (ATTB) and the integrase catalyzes site-specific recombination between the two chromosomes. Excision essentially reverses this process, reforming the ATTP sequence. P, P*, B, and B* denote sequences adjacent to the site of recombination in the phage and bacterial chromosomes, respectively. The numbers represent genes in the phage chromosome.

phages or increase virocell fitness in other ways (see "What's in It for the Bacterium?" on page 261).

Independent Prophages

As usual, not all phages follow the same script. A minority of temperate phages don't turn over control of their replication to the host, nor do they integrate their chromosome. Instead, they take responsibility for their own replication and for getting the daughter prophages into both cells each time the virocell divides. In these regards, their prophages behave like plasmids–independently-replicating, mobile genetic elements found in many Bacteria. Plasmids control their own replication and their copy number, although most exploit host components to do the work of replicating their DNA. Smaller plasmids maintain many copies of themselves in the cell, so many copies that it is certain that each daughter cell will inherit at least one. Larger plasmids and these "plasmid prophages" maintain only a few copies and provide their own mechanisms to partition them between both daughter cells.

Independence (N15)[7] is one of the phages that opted long ago for a plasmid prophage. Most plasmids are circular molecules of dsDNA. Likewise, although many phage chromosomes are linear inside the virion, upon arrival in a host most of them immediately recircularize, and for good reasons. For one, using a linear dsDNA molecule complicates chromosome replication because the typical DNA polymerase cannot replicate such a molecule all the way to both ends. For another, a linear DNA molecule would be attacked by the cell's exonucleases (see "Death by Nuclease" on page 64). As with other phages, Independence's chromosome circularizes after arriving in the cell, but then it deviates from the norm (see Figure 92). It transforms into a linear plasmid prophage and maintains as such in the virocell generation after generation. How does it protect the vulnerable ends of its linear chromosome?

The transformation from circular to linear is handled by Independence's innovative enzyme (protelomerase[8]). It cleaves the newly

---

[7] phage N15, a siphovirus similar to Temperance.
[8] protelomerase: the enzyme that generates covalently-closed hairpin ends on prophage plasmids and subsequently assists in their replication. Protelomerase is also found in a few prokaryotes. It is related in name only to eukaryotic telomerase.

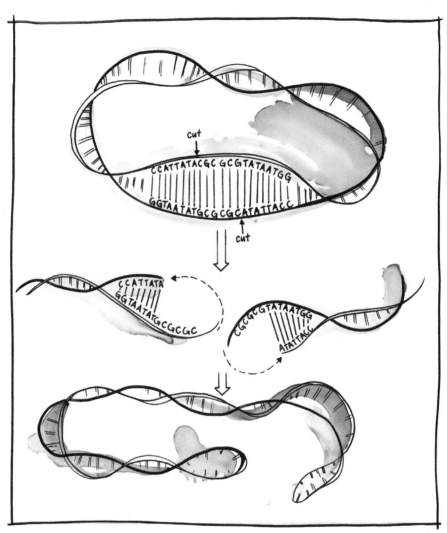

Figure 92: Securing chromosome ends. Soon after arrival, Independence (N15) cleaves its circular dsDNA chromosome by making staggered cuts in both strands. The resulting single-stranded overhangs fold back on themselves and are ligated to their complementary strand, thus forming a hairpin at each end of the linear DNA molecule. This requires both the protelomerase enzyme and the presence of the complementary sequence needed for the base pairing shown in the figure.

formed circular chromosome by making staggered cuts in both strands, which leaves a single-stranded overhang six nucleotides long on each strand. Each overhang folds back on itself and is ligated to the free end of the complementary strand. The net result is a ssDNA hairpin at each chromosome end. This solves one issue–it protects the chromosome ends from exonuclease attack—but it creates another. As you would imagine, such a chromosome structure baffles the host's DNA polymerase. Independence encodes its own enzyme that can replicate the molecule from end to end.

As the lysogen grows and divides, Independence controls its own rate of replication to consistently maintain a ratio of 3-5 prophages to each bacterial chromosome. Importantly, each time the virocell divides, Independence ensures that both daughter cells receive at least one prophage. Without intervention, the prophages would be apt to remain clustered in one region near the site of their birth and would thus all end up together in one of the two daughter cells.

Large plasmids that maintain only a few copies per cell face the same challenge. To meet it, they evolved a number of different mechanisms to actively partition plasmid copies between the daughter cells. In one of their solutions, dynamic filaments connect a centromere-like[9] region in each of the plasmids with one or the other of the cellular chromosomes. When the chromosomes are partitioned to the two daughter cells, their associated plasmids accompany them. The co-existence in cells of these plasmids and one of Independence's ancestors gave Independence all the opportunity it needed to steal the plasmid genes that encode this capability. Subsequent phage generations gradually adapted their chromosome to better interact with those stolen plasmid filaments. A close look at Independence's genome today provides clear evidence of its checkered past—part phage and part plasmid. More than 25 genes certify its close family ties to Temperance and other siphoviruses, while the rest of its chromosome is a mixed bag that includes many genes obviously acquired from plasmids. Why reinvent the wheel when you can co-opt one and adapt it to meet your needs?

Independence deviates from the temperate norm in another respect. Temperance (λ), for example, expresses only a few genes during lysogeny, primarily those required to repress lytic replication and block infection by related phages. It depends on the virocell for its replica-

---

[9] centromere: in eukaryotes, the region of a chromosome that is bound by the spindle filaments to actively partition the chromosomes during cell division. By analogy, the name has been applied to the region of a plasmid or plasmid prophage that binds the filaments that actively facilitate partitioning.

tion and partitioning. Independence's prophage has more work to do. It expresses at least 29 of its 61 genes, including many that assist with its own maintenance generation after generation. Very few temperate phages use plasmid prophages, but Independence demonstrates that it is possible to do this and to do this well. Its prophages are consistently maintained in the cell population, being lost from fewer than one in 20,000 lysogens each virocell generation.

Skinny (Ff) also maintains its chromosomes in the virocell generation after generation without integrating them into the chromosome of its host, but its strategy is strikingly different. Skinny is more accurately described as non-lytic, rather than temperate, because its persistent infection differs from lysogeny in significant ways. It has no need for partitioning machinery because there are enough copies of its independently replicating chromosome that both daughter cells are certain to inherit one. Most, probably all, phage genes are continuously expressed, including those for virion structural proteins. Virions are produced and released continuously without virocell lysis (see Figure 93). Instead of filling preassembled capsids, Skinny's chromosomes acquire their protein coat as they are extruded through the cell membrane (see "Live and Let Live...and Exploit" on page 149). Although the virocell is not killed, the relentless production of so many virions and the membrane damage due to their extrusion takes a toll on the cell. Growth slows markedly.

Till Death Do Us Part

Continuation of the coalition is at the discretion of the prophage. The prophage monitors virocell viability and, when survival is in doubt, it proclaims that it is every chromosome for itself. The prophage terminates the relationship and immediately proceeds with its own replication program – a response referred to as prophage induction.[10] The induced phage's game plan is to assemble a crop of progeny virions while the cell is still capable of supporting production, and then abandon the sinking ship.

---

[10] induction: the termination of lysogeny by activation of the prophage. Phage replication immediately ensues, followed by virocell lysis.

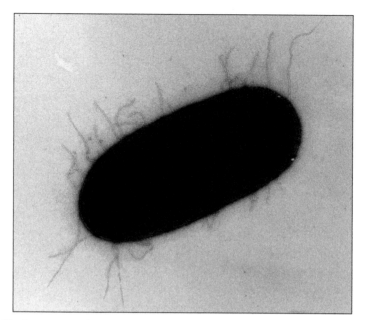

Figure 93: Current production. Skinny's (Ff's) virion production represents a significant drain on host resources and a stress on its CM. This EM image shows an *E. coli* with Skinny's currently extruding virions. Most of them are unfinished, while a few are full length. Credit: Jasna Rakonjac. Reproduced from her doctoral thesis, with permission.

One way that virocell survival can be jeopardized is by damage to its DNA. This is a common occurrence in prokaryotes due to both the UV component of solar radiation and the reactive oxygen species produced by their own metabolism. Their DNA is also harmed by exposure to some antibiotics and other toxins–a risk that is especially prevalent in environments where humans have increased the concentrations of naturally occurring antibiotics or introduced an array of other compounds including synthetic antimicrobials. Of course, Bacteria have evolved multiple mechanisms that detect and repair chromosomal damage, and just as predictably phages have evolved ways to hitchhike on these mechanisms for their own purposes.

The SOS Response

Here I digress from the story of prophage induction and I do so with a specific intent. If you are among the majority who consider Bacteria to be "simple," "primitive" life forms, I wish to give that mindset a little nudge. Although I am unapologetically pro-phage, I do have a great respect for the prokaryotes that not only host the phages, but also are the foundational cellular life forms on Earth—the most ancient, the most diverse, the most versatile. After nearly four billion years of evolution they are highly evolved and admirably expert at what they do. They deserve a place in the spotlight along with the phages.

When Bacteria such as *E. coli* detect DNA damage, they temporarily arrest cell division while they activate a network of genes to make repairs. The initial alert following UV exposure can be sounded by the abnormal generation of regions of ssDNA that are vulnerable to attack by the cell's nucleases. To avoid lethal nuclease strikes, these regions are quickly coated by many copies of a single-stranded DNA-binding protein. These first responders are subsequently replaced by a recombinase (RecA), a key protein for DNA repair in Bacteria, as well as in Archaea and Eukarya.[11] When binding to DNA, each RecA protein shields three nucleotides; many of these proteins polymerize into long filaments that spiral around the chromosome and protect the entire region of ssDNA.

RecA also plays a second role in the bacterium's response to DNA damage. It launches the cell's SOS response[12] by turning on the many genes (approximately 50 in *E. coli*) of the SOS regulon[13] that are scattered across the chromosome. Together these genes effectively deal with the emergency while repairs are made. Normally they are all silenced by a repressor protein[14] that is continuously synthesized by the cell. However, when RecA filaments form on the DNA, this repressor embeds deeply into the spiral groove formed by the filaments. This interaction alters the repressor's conformation, which prompts it to cleave itself and thereby lift the repression of the SOS regulon. Thus, DNA damage leads

---

[11] The RecA homologs in Eukarya and Archaea are Rad51 and RadA, respectively.

[12] SOS response: a coordinated prokaryote response to DNA damage. Gene transcription is altered, DNA repair mechanisms are activated, and cellular priorities are adjusted. If repair is successful, cell metabolism then reverts to normal.

[13] regulon: a group of nonadjacent genes that are coordinately activated or repressed by the same regulatory element.

[14] LexA

to a cascading chain of events: RecA filaments assemble on regions of ssDNA, the filaments interact with the repressor of the SOS response, the repressor self-destructs, and the demise of the repressor allows expression of all the genes for the SOS response. When the DNA has been repaired, the RecA filaments disassemble. The newly synthesized repressor resumes silencing the genes of the SOS response, and cellular activity returns to normal.

A prophage needs to "know" when the virocell is in serious trouble, but how does a prophage – merely an island within the host chromosome – detect this? Many take advantage of the host's own SOS response mechanism to alert them. Moreover, some also use the host's response to initiate their own rapid response to the emergency. During lysogeny, Temperance's prophage suppresses lytic replication by the continuous synthesis of its own repressor protein[15] that silences the genes needed exclusively for the lytic pathway. Like the cellular SOS repressor, Temperance's repressor self-destructs when it encounters RecA filaments. Thus, the same event that activates the SOS response in Temperance's virocell also triggers prophage induction.

Numerous other phages rely on the cell's SOS response for their cue, but, as usual, not all of them do this in quite the same way. For example, one phage[16] uses the cell's SOS repressor to also repress its own lytic replication. This eliminates the need to encode its own repressor and makes for a particularly stable partnership. This tactic also ensures that as soon as the virocell initiates an SOS response, the phage responds in step with immediate induction. Prophage induction in response to DNA damage is so widespread among the temperate phages that researchers routinely use this tactic to detect prophages. Researchers deliberately treat bacterial cultures with agents that damage the virocell DNA. The antibiotic mitomycin C, for example, introduces cross-links into the DNA, which effectively block DNA replication and trigger this emergency response.

---

[15] protein cI
[16] phage GIL01, a temperate phage that infects the Gram-positive insect pathogen *Bacillus thuringiensis*.

Bet Hedging

In actuality, the phage strategy for maintaining lysogeny is even more subtle and more adjustable. For phages like Temperance that rely on their own repressor protein, continuous repression of lytic replication depends on the continual presence of sufficient copies of that repressor to ensure that one is always bound to its chromosomal target. Exactly how many of these repressors should a phage synthesize? The number depends on many factors, including repressor half-life and how apt they are to dissociate from their binding site. Half-life, in turn, is affected by the presence of any RecA filaments even if the amount is insufficient to prompt an SOS response. Low levels of RecA may be present at any time because RecA is also needed to resolve difficulties commonly encountered during DNA replication. Any RecA filaments present may cause some phage repressors to self-destruct. The more repressor proteins synthesized by the prophage, the greater the likelihood that repression will be continuously maintained. Make fewer copies, and a prophage will more often escape repression and "spontaneously" excise. Cellular levels of RecA vary with conditions and chance. Interplay of all of these fuzzy factors results in repressor levels occasionally dropping so low that the prophage is induced. However, this occurs routinely in only a small number of lysogens. How many? In two examples, one prophage per 7,000 to 20,000 lysogens each cell generation.

A phage could further reduce the frequency of these spontaneous events by making more repressor proteins, but would that be beneficial? Maintaining a low frequency of induction ensures that there are always a few virions present in the environment to respond immediately to infection opportunities. Launching some new infections is beneficial for the phage because this is the route for spreading into non-lysogenic host clones. A low level of ongoing spontaneous induction can also benefit the lysogens. The DNA filaments spewed into the environment along with the virions by cell lysis support the growth and maintenance of bacterial biofilms.

Archaeal Phages Do It, Too

The little that is known about the extremophilic archaeal phages confirms that these phages follow similar strategies but employ different tactics. The archaeal genus *Sulfolobus* inhabits seeming uninhabitable environments – pH 1 to 3, temperatures up to 90° C – but even here they can't escape the stalwart phages. These Archaea are infected by members of at least four phage families, each of which has a distinctive virion morphology: spindles, droplets, long flexible filaments, and stiff rods. So far, none of these phages are known to be lytic, but only one – one with a spindle-shaped virion – is known to be temperate. The others seem to be maintained in a more or less stable carrier state – a catch-all phrase meaning that we don't really know what they are doing, but virions are produced and released without lysis. The temperate one[17] does encode an integrase that inserts its chromosome into the host chromosome like a typical prophage. Also like a typical prophage, UV treatment leads to virion production. Underlying this observation are numerous anomalies. For one, this prophage maintains a plasmid copy in addition to the integrated prophage. Also, although the prophage's repressor blocks DNA replication, the genes for virion structural proteins continue to be expressed. Lastly, although induction inactivates the repressor and allows virion production, it does not lead to lysis. Perhaps its virions extrude through the cell membrane as seen for other spindle-shaped virions, such as Biped (ATV). After a period of virion production, lysogeny is restored.

Making a Choice

If you are a temperate phage, then each time you infect a new host you must choose between lysis and lysogeny. When should you opt for one, when for the other? If we were faced with making this decision, we might hire a consultant to advise us, to identify the best course for the long-term survival of our lineage. Said consultant would formulate a mathematical model and assign values to parameters such as the number of hosts available, the number of phages competing for those hosts, virion half-life under current conditions, the metabolic ability of the host to support phage replication, and the predicted burst size. These factors can all be reduced to two underlying questions: whether

---

[17] SSV1, spindle-shaped virus 1.

the cell at hand can support production of an adequate number of virions, and what is the likelihood that the progeny produced will find hosts. Temperance answers both of these questions by monitoring a single parameter – the concentration of one phage-encoded protein immediately after infection.

Because its secondary receptors – and the potential sites for its chromosome delivery – are concentrated around the two poles of the rod-shaped host cell, when multiple Temperance virions adsorb to the same cell, all the incoming chromosomes tend to be clustered in those regions. Immediately upon arrival, transcription of every chromosome gets underway. One of the first proteins synthesized is a key regulatory protein.[18] This protein is also an indicator that informs Temperance's decision. The higher the concentration of this protein, the more likely the phage will establish lysogeny. The more phages that are infecting the cell at about the same time, the more of this protein that is synthesized, and thus the higher the probability of lysogeny. This protein concentration alone gives Temperance a good estimation of the relative number of hosts and of competing phages in the neighborhood.

Temperance goes one step farther and also factors in the metabolic state of the host cell. The concentration of that indicator protein depends not only on the amount synthesized, but also on the metabolic well-being of the cell. Bacteria are smaller when starved than when well fed and rapidly growing. Given the same number of infecting phages, indicator concentration will be higher in smaller cells. Moreover, that indicator protein is subject to degradation by a host protease that is also localized at the poles. The amount of protease present decreases in starved cells, which also increases the local indicator concentration. Thus, using this one protein, Temperance assesses host well-being as well as host availability.

In the artificial world of a lab culture vessel containing only Temperance and its *E. coli* host, a low multiplicity of infection[19] (many well-fed hosts, few virions) prompts Temperance to choose lytic replication. As you increase the number of virions relative to hosts, the probability

---

[18] protein CII

[19] multiplicity of infection: in laboratory experiments, the number of virions added per potential host cell in the culture.

of lysogeny increases until eventually all infections lead to lysogeny. When multiple phages infect concurrently, they each assess the situation independently and vote for lysis or lysogeny. This is not a democratic election. It takes only one vote for lysis to send the virocell down that pathway. A dissenter may integrate nevertheless, just in case lytic replication fails.

Studies of simple one phage/one host model systems in the lab are informative, but undoubtedly oversimplify the situation in natural communities. How this phage decision is modulated by the numerous factors present in a real world situation remains unknown. Likewise, the variations on this theme that have been devised by various other phages await investigation.[20] What environmental factors prompt most temperate phages to choose lysogeny? Which factors favor temperate phages in a community over lytic ones, and vice versa? These, and many other questions currently under investigation at the ecosystem level, are grist for PIC.

Are You a Lysogen?

Prophage induction upon treatment with mitomycin C would seem to offer a simple and direct way to determine what percentage of the Bacteria living in any environment are lysogens. All you would need to do is collect a sample of the bacterial community, add some mitomycin C, and determine what percentage of the Bacteria are lysed. In practice, determining who is a lysogen is anything but simple. Some prophages can indeed be induced by mitomycin C treatment. These are the ones whose excision is triggered by chromosome stress. Induction of some others requires different conditions. Following mitomycin C treatment, one can observe the number of virus-like particles released, but it may

---

[20] While this book was in production, a paper was published reporting a novel mechanism for making the lysis-lysogeny decision. This method, observed in phage φ3T, resembles the quorum sensing used by Bacteria to communicate with one another and to coordinate their actions. During infection of its *B. subtilis* host, φ3T produces a short peptide that is secreted into the environment and then taken up by other *B. subtilis* cells. The intracellular concentration of this peptide is thus an indication of the number of cells in the area that were recently infected by φ3T. Moreover, when this peptide is present inside a host cell, it affects incoming φ3T phages. Specifically, it regulates the transcription of a gene involved in φ3T's lysis-lysogeny decision. Higher intracellular peptide levels favor lysogeny over lytic replication. In this manner, φ3T assesses the number of recent infections and bases its lysis-lysogeny choice on that assessment. See Erez et al. 2017 in "Further Reading" on page 268.

be misleading to assume that these are all functional phage virions. Some may be gene transfer agents (see "A Two-Way Street" on page 153) or defective, non-infectious virions. In practice, these methods have yielded highly variable results – even ranging from 0% to 100% in the same environment. Moreover, induction-based methods detect only those prophages that are able to excise, replicate, and lyse their virocell. Over the course of time, prophages can lose one or more of these abilities through mutation. Although a defective prophage may be undetectable by these tests, some of its genes may still be expressed in the lysogen and may significantly alter the cell's phenotype.[21] The prophage may even be able to excise and release progeny given the presence of a helper phage that provides the missing capability. In natural populations such phage-phage interactions are common and complex (see PIC).

Another way to identify lysogens is to sequence the bacterial genome and look for prophage elements. This approach can also reveal how many different prophages are present in that one bacterial strain. Although straightforward in concept, this, too, is not simple in execution. How do you recognize a prophage within a bacterial genome sequence? The clearest demonstration would be to find an island in the chromosome whose sequence matches that of a known phage. However, only ~2,200 different phages have been sequenced – a small fraction of the total number on Earth. Thus, many prophages would be missed by this criterion. One can look instead for characteristic phage genes, such as an integrase gene. Some prophage integrases can be identified because their amino acid sequence is similar to that of known integrase genes. This method, too, is useful, but limited. Divergent integrases would not be recognized (giving false negative results), and phages are not the only mobile genetic elements that encode an integrase (giving false positive results). However, when several phage genes – perhaps a major capsid protein, a terminase, a DNA polymerase, and an integrase – are present and arranged on the chromosome in the order characteristic of phage genomes, a prophage is clearly implicated. In 2015, one of these genomic approaches identified one or more pro-

---

[21] phenotype: observable traits, such as metabolic activities and morphology, that result from the expression under current conditions of a subset of the genes in a cellular genome.

phages in 82% of all sequenced prokaryote genomes. An immense amount of research remains to be done before we have deciphered the ramifications of lysogeny for the ecology and evolution of even one ecosystem (see PIC).

Why Be Temperate?

To be temperate requires a phage to be equipped for two alternative lifestyles: lytic replication and lysogeny. At a minimum, a temperate phage must have more genes. In addition to all those needed to support lytic replication, the phage must encode a system that decides when to choose lysogeny, tools for insertion into the virocell chromosome and to later excise from it, a mechanism to silence specific genes and express others while a prophage, and factors to sense when it is time to abandon ship. Since these components evolved to function in intimate relationship with a specific host, temperate phages are typically limited to a more narrow host range than their lytic counterparts.

Moreover, lysogeny seems to run counter to the imperatives of lytic replication that reward speed, efficiency, and economy. Instead of rapid production of a hundred or more progeny in a few hours, the prophage merely doubles at the slower pace of host cell division. Temperate phages are not renunciants that are foregoing rampant proliferation. The rapid replication and large burst size seen in the lab are not representative of phage life in most environments. Hosts are typically subject to brief feasts and prolonged famines. While experiencing nutrient limitation or out-and-out starvation, cells are in a stationary state, maintaining but not growing. Having the lysogenic option allows a temperate phage to put lytic replication and progeny release on hold until growth resumes.

What else is gained? Choosing lysogeny avoids the risks of intercellular travel. In many environments, UV irradiation from sunlight can inflict lethal damage on the phage chromosome inside a virion. Hunkered down inside the virocell the prophage gains some protective shading. Moreover, if hit by UV, the DNA repair systems of the host will restore the prophage DNA as if it were its own – as it effectively is at this point. Lysogeny may enable a phage population to sustain during periods of low host abundance. The typical phage is highly host-specific, often limited to a specific bacterial species, even a particular strain. Thus, hosts may be rare even when the bacterial population is robust. When a virion does happen upon a potential host, likely one of its siblings has already taken possession and posted a No Vacancy sign (see "What's in It for the Bacterium?" on page 261). Other not-yet-identified factors may also drive the lysogeny choice. Nevertheless, biding time inside a lysogen, with slow or even no replication, may be preferable to fruitless drifting in search of a host.

Does lysogeny offer security? A prophage is only as secure as its viro-cell. Prokaryote populations are controlled from both above and be-low. Bottom-up controls, such as nutrient limitations, often constrain bacterial numbers by limiting their growth rate, and with that the rate of prophage replication. Most of the time, Bacteria must hunker down and make do, growing very slowly or merely maintaining until conditions improve and growth can surge. It behooves a prophage to help its virocell to survive and even thrive. Prophages often carry metabolic genes that help the virocell to cope with adverse conditions, including nutrient limitation. However, whether growing or not, prokaryotes are also subject to top-down control by predation. Bacteria are the mainstay of the diet of heterotrophic[22] protists, and protists are voracious. About 20% of the marine bacterial community, along with any resident prophages, becomes protist food each day. (By contrast, a far larger percentage of Bacteria die from phage lysis every day, including lysogens as well as non-lysogens. See PIC.)

What's in It for the Bacterium?

Are the benefits of lysogeny all for the phage, the costs all borne by the bacterium? Definitely not! True, the bacterium pays the metabolic cost of replicating some extra DNA, often on the order of an additional 1%. Even though this is a small increase, it may matter. It is also true that every lysogen carries a time bomb that can go off at any moment and is very likely to detonate whenever the virocell launches an SOS response. Nevertheless, the many benefits outweigh the costs. These

---

[22] heterotroph: an organism, such as *Homo sapiens*, that cannot obtain its carbon from simple compounds such as CO_2 and instead requires organic carbon compounds for its carbon source ("food"). Some heterotrophs also obtain their energy from these organic compounds, while others can utilize sunlight.

coalitions have a long history, and I would wager that they will continue as long as there is life on Earth.

Benefit #1: Protection from superinfection. Self-serving prophages employ a great variety of tactics to protect their home from superinfection by other phages. For example, if the same type of phage attempts an infection, the resident prophage's repressor protein will block its lytic replication. The lysogeny option is also blocked for such a newcomer. Integration of the resident prophage split the insertion site in the bacterial chromosome (ATTB), thus prevents another prophage from inserting at this same location. As a result, Bacteria could be abundant – even the specific strain that a phage is seeking – but they could all display "No Vacancy" signs.

The prophage repressor can defend the virocell against other phages, as well. Specifically, it is effective against closely related temperate phages. "Closely related" is defined here as any temperate phage whose repressor protein is so similar to that of the resident prophage that the two repressors are functionally interchangeable. In this scenario, the resident prophage's repressor effectively suppresses lytic replication by the invader. However, if the newcomer's prophage and the resident prophage occupy different locations in the virocell chromosome, the newcomer may proceed to insert at its own specific site. This type of superinfection by a related temperate phage could have led to the multiple prophages found in many bacterial chromosomes today.

Repressor-antirepressor wars can become complicated. Some incoming phages, wary of confronting a repressor on arrival, arrive equipped with an antirepressor. This can counter the resident prophage's repressor and allow the newcomer to launch a lytic infection. However, arriving with an antirepressor would induce any closely related resident prophages. The virocell then becomes home to two or more competing replicating phages. Concurrent replication provides opportunities for recombination between different phages – a mechanism for the exchange of genes or whole gene modules between unrelated phages that share the same host (see PIC).

Other tactics employed to deter superinfection include preventing a superinfecting phage from delivering its chromosome. Some resident prophages modify the structures on the cell surface that serve as phage receptors. When done skillfully, the LPS or pili, for example, can be modified to fool the phages with little loss of fitness for the virocell. Alternatively, the prophage may strike later, attacking after the incoming phage has arrived. Many such tactics remain to be investigated.

Benefit #2: Bacterial diversity. The everyday diversity that we are familiar with in various animal species, humans included, is due primarily to individuals carrying different variants, i.e., alleles, of the same genes. In humans there are, for example, alleles for blue eye color and brown, and for type A or type B antigens on our red blood cells. We don't find some humans with 10% or 20% more genes than others. By contrast, we do find marked differences in gene content from one individual to the next in many bacterial species. How much difference is found and in how many species depends, in part, on how you define a prokaryote species – not a simple matter (see PIC). Typically a bacterial species comprises many strains that all share a set of genes called their core genome. This core is supplemented by a highly variable number of other genes that are present in some, but not all, strains. These are often genes that are useful only under particular conditions. The sum of all the genes found in at least one strain make up the pangenome for that species. Prophages are a major source of these specialization genes present in some strains, but not others.

Benefit#3: Weaponry. Nearly all pathogenic Bacteria carry at least one prophage. Since these prophages are found in only some strains of a species, their genes are part of the pangenome of that species. It is these prophages that encode the toxins and other virulence factors that are essential for the pathogenesis of *Vibrio cholerae*, *E. coli* O157, *Shigella dysenteriae*, *Staphylococcus aureus*, and *Corynebacterium diphtheriae*, among others. A dramatic example of this is the infamous pathogenic *E. coli* O157:H7 strain Sakai. Embedded in its chromosome are 18 prophages and prophage-like elements. Combined they comprise 16% of the bacterial genome and are responsible for much of this bacterium's clout. Similarly, the 4–6 prophages found in *Streptococcus pyogenes* make up 12% of its genome. These are long-standing, intimately coordinated, prophage-host coalitions that increase infection success for these pathogens. Regulation of the expression of some of these genes, such as the diphtheria toxin gene, has been taken over by the host cell to better serve its own interests.

Benefit #4: Competitive adaptations. Prophages typically carry auxiliary genes that are not essential for their own replication but that do enhance survival prospects for the virocell. These genes may facilitate acquisition of phosphate and other nutrients, provide antibiotic resistance, enable detoxification of environmental compounds, and so on. Lysogens may be able to colonize additional ecological niches or to compete more effectively under some conditions. Even in the lab, lysogens are seen to outgrow nonlysogen siblings under some circumstances.

Benefit #5: Self-defense. Some toxins encoded by prophages may protect their hosts from voracious protist predators. For example, one of the toxins that make *E. coli* O157:H7 such a nasty human pathogen also increases the survival of these Bacteria inside the food vacuoles of a protist predator. If the *E. coli* manage to survive long enough, they will be expelled, undigested, from the vacuoles. Seen in this light, human disease is merely unintended collateral damage.

Benefit #6: Horizontal gene transfer. Transduction,[23] i.e., the conveyance of cellular genes from host to host by phages, is a key vehicle for horizontal gene transfer between prokaryotes. This gene trafficking accelerates prokaryote evolution and affects their ecology (see PIC). Successful transfers require a sequence of three low probability events: packaging of the host gene inside a virion, delivery of the packaged DNA into a new host by that virion, and integration of the transferred host gene into the chromosome of the recipient cell by recombination. Successful transduction is indeed a rare event, but irrefutable evidence of its occurrence is evident in prokaryote genomes. There are two ways that host genes can be packaged erroneously inside a virion. One way is the result of imperfect excision of a prophage from the virocell chromosome. Here, the DNA cuts are not made at the exact borders of the prophage, but rather a short distance away. As a result, the excised "prophage" DNA also includes some adjacent host DNA. That bit of

---

[23] transduction: the conveyance of cellular genes between hosts by phage virions and their subsequent incorporation into the recipient cell's DNA by recombination.

host DNA will be packaged in the virion as part of the "phage" chromosome. Since prophages insert at specific locations in the host chromosome, only the subset of host genes that are adjacent to these sites can be transferred by this specialized transduction.[24] Obviously, only a temperate phage whose prophage integrates into the host chromosome can transport host genes this way.

By contrast, any replicating phage can carry out generalized transduction[25] by packaging essentially any region of the host chromosome in a virion. This can happen when the terminase mistakes a similar sequence in the host DNA for the phage's DNA packaging sequence. Some terminases are more error-prone than others, thus making some phages more adept at this type of transduction. Also, for phages that cleave the host chromosome as part of their takeover strategy, the chromosome ends that result can also entice the terminase to package the fragment.

Benefit #7: Donated genes. From an evolutionary perspective, prophages are relatively short-term residents. They insert and excise frequently. However, when a prophage loses its ability to excise due to mutation, its genes are bequeathed to the host cell to use or discard as it sees fit. Most such genes suffer additional mutations and are eventually eliminated from the chromosome, but some have been retained and further evolved to serve the cell. Such co-opted phage genes live on as tailocins, nano-compartments, gene transfer agents, and secretory structures, to name a few (see "A Two-Way Street" on page 153 and "Theft of Intellectual Property" on page 236).

Sporulation and a Savvy Prophage

For billions of years Bacteria have been routinely plagued by hard times including famine, desiccation, and other severe environmental stresses. A few Gram-positive species,[26] including soil-dwelling *B. subtilis*, cope with starvation by sporulating. When a cell commits to this path, it divides into two unequal cells, each of which then follows its

---

[24] specialized transduction: transduction in which genes adjacent to a prophage can be packaged along with the phage chromosome due to imprecise prophage excision.

[25] generalized transduction: transduction in which essentially any region of the host chromosome can be packaged in a virion due to an error by the terminase.

[26] some members of the bacterial phylum Firmicutes.

own distinct developmental path. Transcription is governed by different sigma factors in the two cells, the result being that different groups
of genes are transcribed in each. The larger cell engulfs the smaller,
which then differentiates within the larger mother cell into a dormant
spore. During its maturation, the spore is encased within a tough, protective spore coat and then released by lysis of the mother cell.

Suppose the sporulating cell is a lysogen, specifically a cell of *B. subtilis* strain 168 carrying an SPβ prophage – an intimately intertwined
coalition. During normal growth, the virocell enjoys several benefits
provided by the prophage. Not only does the prophage provide the
usual protection from superinfection by related phages, but it also confers immunity against infection by some lytic phages. Further, it produces an antimicrobial effective against some Gram-positive Bacteria
that lack this prophage. It politely integrates into a gene that is active
only during sporulation and only in the mother cell. When the cell is
growing normally, interruption of the gene by the prophage does not
interfere at all. The protein product of this gene is required only for
synthesis of the polysaccharide of the spore coat. Its presence in that
coat makes the spores hydrophilic, a property that helps them to disperse via water flow in the soil.

Like many other prophages, this one is induced by DNA damage to
the virocell. Virion production and cell lysis quickly follow. In addition, sporulation itself triggers excision of the prophage, but only in
the mother cell. Thus, activity of the interrupted gene is restored in the
cell when and where it is needed to produce a useful spore coat component. Although the prophage excises in the mother cell, no virions
are produced there. Apparently, when associated with the sporulation
sigma factor, the RNAPs ignore the promoters of the phage genes and
thus the genes required for phage replication are not expressed. Meanwhile, the prophage in the spore-to-be stays put in the chromosome
and is inherited by the spore.

A Discriminating CRISPR System

Do you remember the CRISPR defense system that recognizes a previously encountered phage and cleaves the invading

DNA (see "CRISPR Surveillance" on page 75)? Bacterial CRIS-PRs are equal opportunity killers that are just as quick to attack an incoming temperate phage as a lytic one. Do Bacteria have to forego the benefits of lysogeny in order to have a CRISPR defense? Often, yes. If they acquired a spacer targeting a temperate phage, the CRISPR machinery would attack that phage DNA whether it was in an invading phage chromosome or an integrated prophage. Cleaving a prophage would be suicide, as this would cleave the virocell chromosome. But CRISPR systems are not all the same. One type initially tolerates an incoming phage chromosome and also a quiescent prophage. Only when that invader begins transcription or the prophage goes lytic and rapidly transcribes its genes does the CRISPR system strike.[27] These savvy Bacteria enjoy the services of a resident prophage as long as it is well-behaved, but are poised to strike if the prophage threatens the survival of the cell.

Why Not?

If lysogeny is so great for both the phage and the bacterium, we have to wonder why all phages aren't temperate and why all Bacteria aren't lysogens. There are no definitive answers yet, only speculations. Perhaps the lytic life cycle evolved first, with the more complex temperate lifestyle following later. These relative latecomers then, over evolutionary time, came to account for a growing share of the global phage population. This seems likely given that becoming temperate required the evolution of numerous capabilities over and above the basics for lytic replication. Will an even greater proportion of the phage community be temperate some day, or is today's proportion already optimal? In the oceans, the environment for which we have the most data, roughly half of the Bacteria are lysogens, and thus are shielded to some extent from lytic infection. More than half lysogens would further reduce host availability. Without a minimal frequency of lytic infection, a phage population may be driven to extinction.[28]

---

[27] Type III CRISPR/Cas in *Staphylococcus epidermidis*. See Goldberg et al., 2014, in "Further Reading" on page 268.

[28] Drawn from Paul, 2008. See "Further Reading" on page 268.

Extinction? Who cares? Who needs them anyway? We do. Phages – both temperate and lytic – are essential for the well-being of prokaryotes, short-term as well as long-term (see PIC). And the prokaryotes run our world.

Apologia

In this book, as in most introductions to phages, I explored the lytic life cycle first. I dedicated six chapters to the lytic agenda and then, in what could appear to be an afterthought, I tacked on one chapter about lysogeny. Was this fair? Justifiable? There are reasons for this skewed allotment. The replicative life cycle, usually culminating in lysis, is a *sine qua non* for being a phage. It is also the context required for appreciation of the lysogenic phase. Moreover, researchers have justifiably focused on what could be productively investigated with the tools and methods available. In past decades, this focus led to myriad reports about the core steps of infection, replication, and release. Moreover, lytic replication fit comfortably within the pathogenic framework of microbiology of that era. Even today, viruses remain best known for causing disease or death in plants, animals, and Bacteria.

Today's perspective on the world of phage is markedly different. Dramatic methodological advances have engendered increased appreciation and understanding of the importance of microbial ecology – and of phages in particular. We now have eyes with which to see complex communities wherein cooperation is as profuse as is competition. New tools have revealed the prevalence of temperate phages and lysogeny, and the ramifications are still being investigated. I welcome the day when the number of book pages devoted to lysogeny routinely equal, or even surpass, those dedicated to the lytic pathway. For now, I promise more on lysogeny in PIC.

Further Reading

Abe, K, Y Kawano, K Iwamoto, K Arai, Y Maruyama, P Eichenberger, T Sato. 2014. Developmentally-regulated excision of the SPβ prophage reconstitutes a gene required for spore envelope maturation in *Bacillus subtilis*. PLoS Genet 10:e1004636.

Bondy-Denomy, J, AR Davidson. 2014. When a virus is not a parasite: The beneficial effects of prophages on bacterial fitness. J Microbiol 52:235-242.

Brussow, H, C Canchaya, WD Hardt. 2004. Phages and the evolution of bacterial pathogens: From genomic rearrangements to lysogenic conversion. Microbiol Mol Biol Rev 68:560-602.

Butala, M, D Žgur-Bertok, SJ Busby. 2009. The bacterial LexA transcriptional repressor. Cell Mol Life Sci 66:82-93.

Canchaya, C, G Fournous, H Brüssow. 2004. The impact of prophages on bacterial chromosomes. Mol Microbiol 53:9-18.

Erez, Z, I Steinberger-Levy, M Shamir, et al. 2017. Communication between viruses guides lysis–lysogeny decisions. Nature 541:488-493.

Feiner. 2015. A new perspective on lysogeny. Nat Rev Micro 13:641-650.

Fineran, P, N Petty, G Salmond. 2009. Transduction: Host DNA transfer by bacteriophages. in *Encyclopedia of Microbiology*: Elsevier. p. 666-679.

Goldberg, GW, W Jiang, D Bikard, LA Marraffini. 2014. Conditional tolerance of temperate phages via transcription-dependent CRISPR-Cas targeting. Nature 514:633-637.

Golding, I. 2011. Decision making in living cells: Lessons from a simple system. Ann Rev Biophys 40:63-80.

Hendrix, R. 2008. Cell architecture comes to phage biology. Mol Microbiol 68:1077-1078.

Lwoff, A. 1953. Lysogeny. Bacteriological Reviews 17:269-337.

Paul, JH. 2008. Prophages in marine bacteria: Dangerous molecular time bombs or the key to survival in the seas? ISME J. 2:579-589.

Prangishvili, D, K Stedman, W Zillig. 2001. Viruses of the extremely thermophilic archaeon *Sulfolobus*. Trends Microbiol 9:39-43.

Ravin, NV. 2011. N15: The linear phage–plasmid. Plasmid 65:102-109.

Ravin, NV. 2015. Replication and maintenance of linear phage-plasmid N15. Microbiol Spectr 3: PLAS-0032-2014.

Epilogue

Figure 94. A starry night. One-eighth of a microliter of seawater from a coral reef at Guam. The sample was treated with a fluorescent stain (Sybr Gold) that is used to visualize nucleic acids (DNA and RNA) under an epifluorescence microscope. Large objects are prokaryotes and protists, while the smaller dots are virus-like particles, likely virions. Imaged at ~800X magnification. Courtesy of Ben Knowles, Rutgers University.

The phage thoughts that filled these pages were mostly garnered by eavesdropping on phages placed in artificial situations in the lab. Here, large numbers of virtually identical phages are presented with large numbers of virtually identical well-fed hosts. So provisioned, the phages cooperate and reliably perform for us, generation after generation. This gross simplification compared to natural environments enabled researchers to tease out very specific phage thoughts.

Phages think many more thoughts when confronted with the real world (just like us). A phage must deal with many different situations. Some environmental factors may be stable, while others fluctuate dramatically, or a virion may drift from one microenvironment to another. A new virion that one minute is surveying its prospects in someone's gut, may in the next be expelled into a chilly, nutrient poor environment where hosts are scarce and struggling. Even within a bustling prokaryote crowd, of the thousands present only a few may be potential hosts. They, too, are unreliable, as small variations between strains can transform phage success into failure. Competition is fierce. Not only kin, but ten or more different phages may be vying for possession of the same cell. Some may already have staked a claim.

What are the effects of 10^{31} phages thinking hour after hour, year after year, epoch after epoch? Phages evolve rapidly – overnight in a laboratory flask. Their evolution does not simply plod along with one point mutation after another, but rather it also employs creative gene swapping, recombination, inversion, duplication, co-optation, and more. With 10^{31} phage chromosomes participating, even useful combinations of several rare events occur frequently. Natural selection acts just as swiftly. Consider the possibilities!

The 10^{30} prokaryotes are said to run the world, managing all of Earth's biogeochemical cycles as they do. By selectively preying on them and contributing genes to them, the 10^{31} phages control the prokaryote populations, maintain their diversity, and influence their evolution. Who, then actually runs the world? The planned companion book, *Phage in Community*, will offer a phage perspective on that question.

Other Resources

Phage Hunter Programs

PHIRE: http://www.hatfull.org/courses/

> The Phage Hunters Integrating Research and Education (PHIRE) program was established in 2002, with the support of the Howard Hughes Medical Institute (HHMI), by Graham Hatfull of the University of Pittsburgh to provide research lab opportunities for local high-school and university students. Over the years, the participating students have isolated, named, sequenced, and annotated a thousand new phages from their soil samples – real scientific research!

SEA-PHAGES: http://seaphages.org/

> PHIRE expanded to the classroom setting as the HHMI Science Education Alliance-Phage Hunters Advancing Genomics and Evolutionary Science (SEA-PHAGES) program. In this form, the phage discovery experience now reaches students at 125 participating institutions. The typical program is a two-semester, discovery-based undergraduate research course that begins with students collecting soil samples and proceeds through virus isolation to lastly genome annotation and bioinformatic analyses of their newly discovered phages.

Books

Calendar, R. 1988. *The Bacteriophages*. Vol. 1: Plenum.

Calendar, R. 1988. *The Bacteriophages*. Vol. 2: Plenum.

Calendar, R. 2006. *The Bacteriophages* (2nd edition). Oxford University Press on Demand.

Cann, AJ. 2001. *Principles of Molecular Virology* (standard edition). Academic Press.

Mateu, MG. 2013. *Structure and Physics of Viruses: An Integrated Textbook*. Springer Science & Business Media.

Ptashne, M. 1993. *A Genetic Switch* (2nd edition). Blackwell Scientific.

Rohwer, F, M Youle, H Maughan, N Hisakawa. 2014. *Life in Our Phage World*. Wholon.

Roossinck, MJ. 2016. *Virus: An Illustrated Guide to 101 Incredible Microbes*. Princeton University Press.

Rossmann, MG, VB Rao. 2011. *Viral Molecular Machines*. Springer Science & Business Media.

Witzany, G. 2012. *Viruses: Essential Agents of Life*. Springer Science & Business Media.

Note: The many popular science books available about viruses generally focus on the viruses that infect eukaryotes, and most especially on those that cause disease in us or our domesticates. Similarly, the respected virology textbooks devote few pages to phages, or even to the viruses that infect microbial eukaryotes.

Online Resources

2015 Year of the Phage: 2015phage.com

> companion site to the 2015 Year of the Phage Meeting in San Diego, CA. Here you'll find 30 informal talks by phage researchers and phage art from the conference, as well as links for downloading a free PDF file of *Life in Our Phage World* (see book list).

EMBO Conference – Viruses of Microbes 2012: http://bit.ly/2iQFYpK

> 17 conference presentations via YouTube, many of which cover topics that are directly relevant to this book, while others discuss the obstacles impeding the clinical application of phage therapy.

IBioSeminars: ibiology.org/ibioseminars.html

> a large and growing collection of accessible seminars given by leading researchers in diverse biological realms, including viruses, evolution, nucleic acid structure, and protein translation. Explore freely and you might be delighted by stories on unrelated topics such as cephalopod (squid) signaling and camouflage.

Small Things Considered: http://schaechter.asmblog.org/

> the flagship blog of the American Society of Microbiology. In addition to ongoing posts, the archives offer access to more than ten years of biweekly posts, including a significant number about phages and other viruses – one small part of the vast breadth of the microbiological world explored here.

The Bacteriophages: TheBacteriophages.org

> companion site to *The Bacteriophages* (Richard Calendar, ed.) that displays numerous informative illustrations of phages and the phage life cycle.

The Viral Zone: http://viralzone.expasy.org/

> some basic viral molecular biology combined with a characterization of every viral family including excellent schematic drawings of their virion architecture. This is a service of the SIB Swiss Institute of Bioinformatics.

This Week in Virology (TWiV): http://www.microbe.tv/twiv/archive/

> popular weekly podcasts hosted by Vincent Racaniello that feature informal discussions about recent virus research or related news. Be forewarned that the spotlight is on pathogenic viruses of plants and animals ("the kind that make you sick"), with only an occasional digression into the world of phage.

The Virology Blog: http://www.virology.ws/

> another offering by the prolific Vincent Racaniello (see above). Here, too, the emphasis is on eukaryotic viruses, but all viruses share many of the same strategies.

Glossary

Acidianus Two-tailed Virus: (Biped) a phage that infects *Acidianus convivator* (an acid-loving, hyperthermophilic crenarchaeon) and that is noted for growing two long tails after leaving its virocell.

adaptive immunity: a form of immunity long thought to be limited to the vertebrates that defends against specific, previously-encountered pathogens.

Alc protein: one of the proteins made by **Lander** during the first two minutes after arrival. It acts as a transcription terminator that halts transcription at specific C-containing sequences, thus selectively interfering with transcription of host genes.

Alt protein: an **internal protein** of **Lander**, about 40 copies of which are delivered during infection along with its DNA chromosome.

Archaea: one of the three domains of life along with the **Bacteria** and **Eukarya**; also classified as a **prokaryote**.

arginine: one of the basic amino acids, thus one that typically carries a positive charge *in vivo*.

autolysin: an enzyme that digests an organism's own cells. In prokaryotes, specifically a **murein** or **pseudomurein** hydrolase that cleaves specific bonds in **peptidoglycan**.

autotroph: an organism, such as a photosynthetic bacterium, that can obtain its carbon from simple inorganic compounds, e.g., CO_2.

auxiliary metabolic gene: (AMG) phage genes that function in cellular metabolism. Initially they were thought to be the exclusive property of cells and only incidentally associated with phage.

Bacillus phage φ29:   (Dynamo) a **podovirus** who uses the enzymatic activity of its short tail to nibble a pathway through the cell wall surrounding its Gram-positive host.

Bacteria: one of the three domains of life along with the **Archaea** and **Eukarya**; also classified as **prokaryotes**.

Bacteriophage SPP1:   (Positivist) a **siphovirus** who uses as its receptor a cellular protein that extends outward beyond the thick cell wall of its Gram-positive host.

biofilm: a structured community of microbes that adhere to each other and also typically to a surface. Often they live embedded within a matrix of secreted extracellular polysaccharides combined with proteins and DNA. The matrix provides protection while maintaining the structure that facilitates cooperative interchanges between cells.

Biped: *see Acidianus* **Two-tailed Virus.**

Bordetella phage BPP-1:   (Fickle) a **podovirus** who uses its sophisticated **diversity generating element** to specifically alter its **receptor binding protein.**

capsid: the protein shell that surrounds the phage chromosome(s) within the **virion.**

capsomer: a subunit of a phage **capsid** comprising a hexameric or pentameric ring of capsid proteins. Often capsomers pre-assemble from individual capsid proteins, then self-assemble with other capsomers (and sometimes a portal) to form the capsid.

Caudovirales: the viral order that comprises the vast majority of characterized phages and that includes three families: *Myoviridae, Siphoviridae,* and *Podoviridae.*

Caulobacter phage φCbK: (Nerd) a **myovirus** with an unusual head filament that it uses to attach to a host **flagellum** during infection.

cell wall: the protective covering completely surrounding the cell membrane of **prokaryotic** cells and composed of multiple layers of **peptidoglycan.**

centromere: in eukaryotes, the region of a chromosome where the spindle filaments attach to actively partition the chromosomes during cell division. By analogy, the name has been applied to the region of a **plasmid** or **plasmid prophage** where the filaments attach to actively facilitate partitioning.

chaperone: a protein that assists the correct folding of other protein molecules or recognizes misfolded proteins and helps them to refold correctly.

Chimera: *see* **Enterobacteria phage P22.**

clone: a group of cells or organisms, such as **Bacteria** or **Archaea,** that are descended from a single ancestor and that are genetically identical.

CM: Cell membrane.

cognate: (adj.) a term borrowed from linguistics and used to indicate some correspondence between two molecules. For example, a **DNA methyltransferase** and its cognate **restriction endonuclease** recognize the same methylation pattern in the same **palindrome.**

coliphage: a phage that infects the bacterium *Escherichia coli.*

coliphage φX174:   (Yoda) a **microvirus** who overlaps genes to encode more proteins in its very small, circular ssDNA chromosome.

competence: in **Bacteria**, the ability to take up DNA from the environment. Most often this uptake is activated during starvation and the incoming DNA is recycled. However, some Bacteria selectively take up DNA from close kin and occasionally **recombine** the fragments into their own chromosome.

conjugation: a form of bacterial "sex" in which DNA is transferred from one bacterium to another through cell-to-cell contact. Often the process is coordinated by a conjugative **plasmid** that encodes the necessary proteins.

conspecific: a member of the same species.

Corticoviridae: a bacteriophage family whose circular dsDNA chromosomes travel inside icosahedral **virions** that assemble with an internal membrane.

Cowboy: *see Salmonella* **phage χ**.

Crenarchaeota: a major phylum within the domain **Archaea** that includes many **hyperthermophiles**.

CRISPR: (**C**lustered **R**egularly **I**nterspaced **S**hort **P**alindromic **R**epeat) a **prokaryotic** form of adaptive immunity that involves the memory, recognition, and cleavage of previously encountered invading nucleic acids, such as phage chromosomes.

cryo-EM: a variant of **transmission electron microscopy** in which samples are prepared by rapid freezing, often by plunging into liquid ethane, and then imaged without other chemical treatments, such as staining.

cyanophage: a phage that infects a member of the Cyanobacteria (a phylum of photosynthetic bacteria).

Cystoviridae: a bacteriophage family whose **segmented** dsRNA chromosomes travel inside icosahedral **virions** composed of inner and outer **capsids** and that assemble with an external membrane.

decoration protein: a **structural protein** added to a capsid after assembly that may reinforce the **capsid**, alter the overall electrical charge of the capsid, facilitate binding to various surfaces, or serve other purposes.

dimer: a complex formed by the association of two **macromolecules**, typically proteins, that are usually non-covalently joined.

diversity generating retroelement: a cluster of genes that introduces mutations in precise locations in a target gene, while maintaining the ability to restore the original sequence from an unaltered template copy. The gene cluster includes a **reverse transcriptase** that participates in the mutagenesis step.

DNA methyltransferase: an enzyme that transfers **methyl groups** from a donor molecule to specific bases in DNA.

DNA polymerase: a family of enzymes that catalyze the synthesis of DNA by the polymerization of nucleotides (including proofreading of the newly synthesized strands) and also carry out DNA repair functions.

domain: one of the three major divisions of all cellular life, i.e., the **Bacteria, Archaea,** and **Eukarya**. *See also* **protein domain**.

Dynamo: *see Bacillus* **phage** φ29.

electron microscope: a microscope that images small objects using a beam of electrons rather than visible light to achieve the higher magnification and resolution required to visualize fine details within nanometer-scale biological structures, including **virions**.

emergent property: a property of a group that does not reside in any individual member, but emerges from the combination of the lower level actions of each of the group members.

endocytosis: an active process carried out by all **eukaryotic** cells to acquire molecules or small particles that cannot penetrate the cell membrane. A small region of the cell membrane invaginates and then pinches shut at the neck to form a vesicle in the cytoplasm that contains a small volume of the extracellular fluid.

endolysin: a phage enzyme released from inside the **virocell** that cleaves bonds in **peptidoglycan**, thereby lysing the cell.

endonuclease: a **nuclease** that cuts a nucleic acid strand at an internal location. *See also* **exonuclease**.

enteric: (adj.) pertaining to the intestines, often used to denote **prokaryotes** that are found in the gut of humans or other vertebrates.

Enterobacteria phage Ff: (Skinny) a group of filamentous phages that includes f1, fd, M13, and others, all of whose genome sequences share 98% identity. They all infect *E. coli* through interaction with the F **pilus**.

Enterobacteria phage HK97:   (Lancelot) a **siphovirus** that stabilizes its capsid by interlocking the **capsid** proteins to form resilient chain mail.

Enterobacteria phage λ: (Temperance) the most extensively studied **siphovirus** that provided the prototype for numerous molecular processes including gene regulation, **virocell lysis**, and the mechanism of **lytic/lysogenic** decision making.

Enterobacteria phage N15:   (Independence) a **siphovirus** whose
    prophage does not integrate into the host chromosome, but instead
    replicates in the **virocell** cytoplasm.

Enterobacteria phage P22:   (Chimera) a phage classified by different
    criteria as either a **podovirus** or a **siphovirus** and who carries out
    generalized transduction relatively frequently.

Enterobacteria phage P4:   (Thief) a **podovirus** who steals its **virion**
    proteins and some other components from a co-infecting helper
    phage.

Enterobacteria phage P4:   (Slick) a **tectivirus** who uses the membrane
    inside its icosahedral **virion** to form a DNA delivery chute.

Enterobacteria phage Qβ: (Minimalist) a **levivirus** who carries out
    infection and **lytic replication** with a mere three genes.

Enterobacteria phage T4:   (Lander) a **myovirus** with a large, complex
    virion whose cellular takeover, replication, assembly, and
    chromosome delivery have been intensely studied.

Enterobacteria phage T7:   (Stubby) a **podovirus** who uses a few
    proteins packaged inside its **capsid** to extend its short tail during
    DNA delivery.

Eukarya: one of the three domains of life along with the **Bacteria**
    and **Archaea**, whose intracellular structures include membrane-
    bounded compartments.

Euryarchaeota: a very diverse phylum within the domain **Archaea**
    that includes anaerobic methane producers (methanogens)
    and extreme halophiles (that require a high salt concentration
    environment), as well as some thermophiles (that prefer a warm
    environment).

exonuclease: a **nuclease** that clips off nucleotides from the ends of a
    linear nucleic acid molecule. *See also* **endonuclease**.

extremophile: an organism that thrives in environments with
    extremely high or low temperature, pressure, salinity, or pH
    (acidity or alkalinity).

Fickle: *see Bordetella* **phage BPP-1**.

flagellotropic phage: a phage whose **virion** recognizes and adsorbs to
    a **flagellum** of its potential host.

flagellum (plural, flagella):   an appendage of many **prokaryotic** cells
    that is anchored in the cell membrane and rotates to propel the cell
    through the milieu.

F pilus: a type of sex **pilus** thought to form a conduit for the transfer
    of DNA from one **prokaryotic** cell to another during **conjugation**.

Fusion: *see* **phage PM2.**

gene transfer agent: a virus-like particle that is produced and released by a **prokaryote**, and that contains exclusively cellular DNA.

generalized transduction: **transduction** in which essentially any region of the host chromosome can be packaged in a **virion** due to an error by the **terminase**. *See also* **transduction** and **specialized transduction.**

glycan: a chain composed of multiple sugar molecules that may be an independent molecule or may be attached to another molecule such as a protein or nucleic acid.

glycoprotein: a protein with attached **glycans.**

habergeon: a short, sleeveless shirt of chain mail.

heterodimer: a functional unit composed of two different **macromolecules**, most often proteins, joined by non-covalent bonds.

heterotroph: an organism, such as *Homo sapiens*, that cannot obtain its carbon from simple compounds such as CO_2 and instead requires organic carbon compounds for its carbon source ("food"). Some heterotrophs also obtain their energy from these organic compounds, while others can utilize sunlight.

Hoc protein: (highly immunogenic outer capsid protein) a non-essential **decoration protein** attached to the **capsid** of **Lander.**

holoenzyme: a functional enzyme complex composed of multiple subunits, e.g., the **RNA polymerase** holoenzyme.

homodimer: a functional unit composed of two copies of the same **macromolecule**, most often a protein, joined by non-covalent bonds.

homologous genes: genes present in two different types of phage (or in two different species of Bacteria or Archaea) that were inherited by both from a common ancestor.

homologous recombination: the reciprocal exchange of segments between two DNA (and less frequently RNA) strands that requires the strands to share a region of near identical sequence of at least ~20 bp.

horizontal gene transfer: the movement of one or more genes from the chromosome of one **prokaryote** or conjugational **plasmid** to the chromosome of another cell of the same or a different prokaryote species (or to a **eukaryote**).

hyperthermophile: an organism or virus that thrives at temperatures of 80° C and above.

icosahedron: a geometric solid with twenty flat faces (each face being an equilateral triangle) and twelve vertices.

Independence: *see* **Enterobacteria phage N15.**

induction: the termination of **lysogeny** by activation of the **prophage**. Phage replication immediately ensues, followed by **virocell lysis**.

innate immunity: any inherent defense against invading nucleic acids or pathogens in general, in contrast to **adaptive immunity** that is targeted against specific, previously-encountered pathogens.

Inoviridae: a bacteriophage family whose circular ssDNA chromosomes travel inside filamentous **virions** that extrude from the host cell.

integrase: the enzyme encoded by **temperate** phages that catalyzes insertion of a **prophage** into the host chromosome by **site-specific recombination**. This phage integrase is distinct from the retroviral integrase that acts on DNA produced by **reverse transcription** of an RNA virus and inserts it into the DNA in **eukaryotic** cells.

internal protein: a protein that is packaged inside the **capsid** along with the phage chromosome.

invertible gene cassette: a chromosome segment whose orientation within the chromosome is periodically reversed. An invertase catalyzes the site-specific cutting of the segment from the chromosome followed by its pasting into the same location in inverted orientation. *See also* **site-specific recombination**.

Lancelot: *see* **phage HK97.**

Lander: *see* **Enterobacteria phage T4.**

Lemon-shaped archaeal virus His1: (Spindly) an archaeal virus who packages its DNA chromosome inside a spindle-shaped **virion** that rearranges to form a tube during DNA delivery.

Leviviridae: a bacteriophage family whose ssRNA chromosomes travel inside small icosahedral **virions.**

liposome: a small, membrane-bounded vesicle, typically artificially produced *in vitro*.

liquid crystal: a physical state intermediate between a liquid and a crystal in which a substance may flow like a liquid while its molecules remain in a stable crystalline array.

lysin: any molecule that causes cell lysis, such as phage lysins that cleave bonds in **peptidoglycan**. *See also* **endolysin**.

lysine: one of the basic amino acids, thus one that typically carries a positive charge *in vivo*.

lysis: the rupture of a **prokaryote** cell membrane that results when a weakened or damaged **cell wall** is unable to counter the internal **turgor pressure** of the cell.

lysogen: a **virocell** with one or more resident **prophages**.

lysogeny: a phage life cycle in which **lytic replication** is postponed for an indefinite period of time during which the phage is maintained as a **prophage** within the **virocell**.

lytic replication: the replication of a phage after arrival in a host cell leading quickly to the release of progeny **virions** by host **lysis**.

macromolecule: a molecule containing a very large number of atoms, often assembled by linking smaller molecules termed building blocks. For example, proteins, nucleic acids, carbohydrates, and lipids.

ManY: a protein that is part of the **CM** mannose transporter in *E. coli* that serves as the **secondary receptor** for Temperance.

metabolite: any of the many small molecules that participate in or are produced by cellular metabolism.

methyl group: $-CH_3$, i.e., one carbon atom with three attached hydrogen atoms.

Microviridae: a bacteriophage family whose circular ssDNA chromosomes travel inside small icosahedral **virions**.

Minimalist: *see* **Enterobacteria phage Qβ**.

mobile genetic element: a molecule of DNA that typically encodes one or more proteins and that can move between **prokaryotic** cells or between locations on the chromosome(s) within a cell. These include **plasmids, transposons**, and others.

multiplicity of infection: in laboratory experiments, the number of virions added per potential host cell in the culture.

murein: the specific type of **peptidoglycan** used by **Bacteria**.

Myoviridae: a bacteriophage family whose linear dsDNA chromosomes travel inside **virions** composed of an icosahedral **capsid** and a long, contractile tail.

negative-sense: refers to an RNA or DNA molecule whose complementary sequence encodes the amino acid sequence of a protein. *See also* **positive-sense**.

Nerd: *see Caulobacter* **phage φCbK**.

noncoding RNA: an RNA molecule that does not function as mRNA, i.e., whose sequence is not translated into the amino acid sequence of a protein. Some noncoding RNAs function in translation or regulate cellular activities.

nuclease: an enzyme that cleaves a molecule of DNA or RNA. **Endonucleases** cut internally; **exonucleases** clip off nucleotides from the ends of a linear molecule.

nucleate: (verb) to facilitate the first step in the formation of a structure, such as a **capsid**, by self-assembly of its component parts.

oligomer: (a) a complex formed by the association of a few macromolecules, typically proteins, usually non-covalently bound; (b) a molecule formed by covalent linkage of multiple building blocks but fewer such units than in a **macromolecule**, e.g., a **glycan** composed of six sugars.

OM: The outer membrane of a Gram-negative bacterium.

orphan DNA methyltransferase: a **DNA methyltransferase** that is not accompanied by a **cognate restriction endonuclease** (RE).

palindrome: in a double-stranded section of a DNA or RNA molecule, the presence of the identical sequence in both strands when they are read in opposite directions.

peptide bond: the covalent bond linking consecutive amino acids in a protein chain. The acid portion of one amino acid (a carboxyl group) is joined to the amino portion (the amino group) of the next amino acid.

peptidoglycan: a molecular mesh that forms the **prokaryotic cell wall** and is composed of chains of specific sugars cross-linked by short chains of amino acids (peptides). *See also* **murein** and **pseudomurein**.

phage Ff: *see* **Enterobacteria phage Ff**.

Phage in Community: (PIC) the companion book to this one planned for publication in 2019.

Pharaoh: *see* ***Sulfolobus* turreted icosahedral virus STIV**.

phenotype: observable traits, such as metabolic activities and morphology, that result from the expression under current conditions of a subset of the genes in a cellular genome.

PIC: *see* ***Phage in Community***.

pilus (plural, pili):   straight, filamentous appendages of some **Bacteria** and **Archaea** that carry out one of a variety of functions such as attachment to host cells or other surfaces, transfer of DNA between cells, and motility.

plasmid: an independently replicating DNA molecule, separate from the cell's chromosome(s), that is found in many **prokaryotes**. Plasmids are inherited vertically, but also are sometimes transferred horizontally to another cell by **conjugation**. Some **prophages** do not integrate in the host chromosome, but instead maintain as a plasmid prophage.

PM2:   (Fusion) a **corticovirus** who delivers its chromosome into a host cell by fusing its internal membrane with the cell membrane of its host.

Podoviridae: a bacteriophage family whose linear dsDNA chromosomes travel inside **virions** composed of an icosahedral **capsid** and a short tail.

point mutation: the replacement of one base in a chromosome by another.

polysaccharide: a polymer of monosaccharides, i.e., sugars.

porin: proteins embedded in the outer membrane of Gram-negative **Bacteria** that span the membrane and form pores that allow the free passage by diffusion of selected small molecules.

positive-sense: refers to an RNA or DNA molecule whose sequence encodes the amino acid sequence of a protein. *See also* **negative-sense**.

Positivist: *see* **bacteriophage SPP1**.

primary receptor: the molecular structure on the surface of a cell that serves as the site of the initial, reversible adsorption by a phage **virion**. In some cases it also serves as the **secondary receptor**.

primary structure: the base sequence in a nucleic acid or the amino acid sequence in a protein.

procapsid: a **virion** assembly intermediate, specifically an assembled icosahedral **capsid** prior to chromosome packaging.

prokaryote: a term used to refer to both the **Archaea** and the **Bacteria**, both groups being unicellular organisms that share a cellular architecture that lacks membrane-bounded compartments, thus lack a true nucleus.

prolate: (adj.) elongated; for a sphere (or an icosahedron), being elongated such that the distance between the poles is greater than the diameter at the equator.

promoter: a region in the DNA upstream of the transcription start site that is recognized by the **RNA polymerase holoenzyme** and promotes initiation of transcription of the DNA downstream.

prophage: a phage chromosome that is not actively replicating but is maintained stably in a **virocell**, usually integrated into the cell's chromosome.

protease: an enzyme that cleaves the **peptide bonds** that link one amino acid to another in a polypeptide or protein. Also known as a peptidase or proteinase.

protein domain: a unit of protein structure that can evolve and function independent of other domains in that protein. A multi-domain protein may carry out a coordinated action in which different domains are responsible for distinct functions such as the localization of the protein in the cell, substrate binding, ATP hydrolysis, and catalysis.

protelomerase: the enzyme that generates covalently-closed hairpin ends on **prophage plasmids** and subsequently assists in their replication. Protelomerase is also found in a few **prokaryotes**. It is related in name only to **eukaryotic** telomerase.

protist: an informal term for unicellular **eukaryotes**, an extremely diverse group of both **autotrophs** and **heterotrophs** that comprises ~20 phyla. Familiar protists include amoeba, paramecia, dinoflagellates, diatoms, and algae.

proton motive force: (PMF) a force created when the metabolically-driven proton pump drives protons across the cell membrane to the outside of the cell, thereby producing a higher proton concentration outside the cell. The difference in concentration creates the PMF that pushes protons back in. This push is harnessed by the cell to generate usable chemical energy in the form of ATP.

Pseudomonas phage φ6: (Shy) a **cystovirus** who defends its dsRNA chromosome against **nuclease** attack by keeping it inside a protective **capsid** throughout an infection.

pseudomurein: the specific type of **peptidoglycan** used by **Archaea**.

quaternary structure: (protein) the arrangement of two or more protein molecules that associate through non-covalent bonds. Examples: the **RNA polymerase holoenzyme**, **Lander**'s long tail fibers.

RecA: a bacterial recombinase (an enzyme that catalyzes **recombination**) with homologs in both **Archaea** and **Eukarya**. Its functions include repairing DNA by swapping homologous regions between flawed strands to generate two complete, undamaged chromosomes.

receptor: the molecular structure on the surface of a cell to which a phage **virion** specifically attaches, reversibly or irreversibly, as the first step in infection. *See also* **primary receptor** and **secondary receptor**.

receptor binding protein: the **virion structural protein** that recognizes and binds to the specific structure that serves as the phage's **receptor** on the surface of a potential host cell.

recombination: in prokaryotes and phages, the integration of a segment of exogenous DNA (or RNA) into a DNA (or RNA) chromosome by the enzymatic cutting and rejoining of the DNA (or RNA) molecules. *See* **homologous recombination**.

regulon: a group of nonadjacent genes that are coordinately activated or repressed by the same regulatory element.

replicase: an **RNA-dependent RNA polymerase** (RdRP).

restriction endonuclease: (RE) an **endonuclease** that recognizes specific sites in DNA, typically 4–8 nucleotides long, and cleaves the DNA if the site lacks the correct pattern of base methylation.

reverse transcriptase: an **RNA-dependent DNA polymerase**, i.e., an enzyme that synthesizes complementary DNA using single-stranded RNA as the template.

riboswitch: an RNA molecule that regulates its own synthesis by binding to a specific **metabolite** or other small molecule.

ribozyme: an RNA molecule that functions as an enzyme, i.e., a catalytic RNA.

RNA-dependent RNA polymerase: (RdRP) an enzyme that synthesizes a complementary strand of RNA using an RNA template.

RNAP: *see* **RNA polymerase.**

RNA polymerase: (RNAP) a **holoenzyme** that selectively transcribes regions of a double-stranded DNA chromosome using the base sequence of one DNA strand as its template. It is also known as DNA-dependent RNA polymerase.

Salmonella phage χ: (Cowboy) a **siphovirus** that adsorbs to a host cell by lassoing a rotating **flagellum**.

secondary receptor: the molecular structure on the surface of a cell that serves as the initial site of irreversible adsorption and chromosome entry for a phage. In some cases it is the same as the **primary receptor**.

secondary structure: (nucleic acid) the three-dimensional structure of a DNA or RNA molecule that results from the formation of weak bonds (e.g., hydrogen bonds) between complementary bases within the same molecule or with an associated nucleic acid molecule. Examples: double helix, stem-loop.

secondary structure: (protein) the three-dimensional structure of a segment of a protein molecule resulting from hydrogen bonds between the atoms engaged in the peptide bonds along the backbone. The three most common secondary structures are the α-helix, β-sheet, and random coil.

segmented genome: a genome that is encoded by two or more separate chromosomes.

sex pilus: a pilus that is essential for **conjugation** and that is typically encoded by a conjugative **plasmid**.

Shy: *see* ***Pseudomonas*** **phage ɸ6**.

sigma (σ) factor: the subunit of the **RNAP holoenzyme** that recognizes and facilitates binding of RNAP to specific **promoters**, thereby initiating transcription of the downstream gene(s).

Siphoviridae: a bacteriophage family whose linear dsDNA chromosomes travel inside **virions** composed of an icosahedral **capsid** and a long, flexible, non-contractile tail.

site-specific recombination: a form of **recombination** that occurs at a specific site. It requires a short region of sequence homology between the participating chromosomes and a specific enzyme (e.g., **integrase**) to catalyze cleavage of both chromosomes, their exchange, and their rejoining.

Skinny: *see* **Enterobacteria phage Ff**.

S-layer: a paracrystalline, monomolecular layer of identical proteins that comprises the outermost envelope layer of some **prokaryotic** cells.

Slick: *see* **Enterobacteria phage PRD1**.

specialized transduction: **transduction** in which genes adjacent to a **prophage** can be packaged along with the phage chromosome due to imprecise prophage excision. *See also* **transduction** and **generalized transduction**.

Spindly: *see* **Lemon-shaped archaeal virus His1**.

structural protein: a protein that is a structural component of the **capsid** (or tail or membrane) of a mature virion.

Stubby: *see* **Enterobacteria phage T7**.

Sulfolobus turreted icosahedral virus STIV: (Pharaoh) a virus with an icosahedral **virion** that infects a **hyperthermophilic crenarchaeon** and builds proteinaceous pyramids on the cell surface for virion escape.

superinfection exclusion: various processes whereby a phage within a **virocell** actively blocks infection of that cell by another phage.

surface layer: *see* **S-layer**.

swarm intelligence: the apparently intelligent collective behavior that results from the localized interactions of many individuals within a decentralized, self-organizing system.

tailocin: a co-opted phage tail, encoded by and produced by a **prokaryote**, that upon release from the producer cell targets and kills sensitive cells.

Tectiviridae: a bacteriophage family whose linear dsDNA chromosomes travel inside icosahedral **virions** that assemble with an internal membrane.

teichoic acid: a complex polymer of sugars, an alcohol, and phosphate found in the **cell wall** of Gram-positive **Bacteria**.

TEM: *see* **transmission electron micrograph**.

Temperance: *see* **Enterobacteria phage λ**.

temperate phage: a phage that is capable of two modes of infection: immediate **lytic replication** and **lysogeny**.

tertiary structure: (protein) the overall shape of a correctly-folded, functional protein molecule that is superimposed on the regions of **secondary structure**.

terminase: a motor protein that uses energy from ATP to translocate linear dsDNA into a **procapsid**.

Thief: *see* **Enterobacteria phage P4**.

transcriptional regulation: control of gene expression by regulating the rate of transcription. *See also* **translational regulation**.

transduction: the conveyance of cellular genes between hosts by phage **virions** and their subsequent incorporation into the recipient cell's DNA by **recombination**. *See also* **generalized transduction** and **specialized transduction**.

translational regulation: control of gene expression by regulating the rate of translation or the turnover of the corresponding mRNA. *See also* **transcriptional regulation**.

transmembrane domain: a **protein domain** usually with an α-helical secondary structure that typically contains many hydrophobic amino acids and that readily embeds in lipophilic (lipid loving) membranes.

transmission electron micrograph: (TEM) an image produced by irradiating very small particles or ultra-thin slices of a sample with a beam of electrons. This method can provide higher magnification and resolution than is possible when irradiating with visible light. The differential absorption of electrons due to the varying composition or structure within the sample is sometimes supplemented by staining or shadowing techniques to increase image contrast.

transposon: a diverse family of **mobile genetic elements** that can move from one site to another within or between chromosomes. They are typically short DNA segments that encode only a few proteins in addition to those required for transposition.

turgor pressure: in **prokaryotes**, the pressure that continually pushes the cell membrane against the **cell wall**. This force results from the inward flow of water into the cell driven by osmotic pressure, and is countered by the resistance of the intact cell wall.

virion: the inert, extracellular dispersal form of a phage composed of the phage chromosome(s) and internal proteins surrounded by a protein shell (**capsid**) that sometimes bears a tail, tail fibers, tailspikes, and/or a lipid membrane.

virocell: a host cell that contains an intact phage chromosome that is capable of replication and **virion** production inside this cell.

viroid: an infectious agent that infects plants and that consists of only a circular, single-stranded molecule of **negative-sense** RNA without a **capsid**.

virome: a viral metagenome, i.e., a sequence library obtained by sequencing a sample of DNA or RNA extracted from an entire viral community.

Yoda: *see* **coliphage ϕX174**.

Index

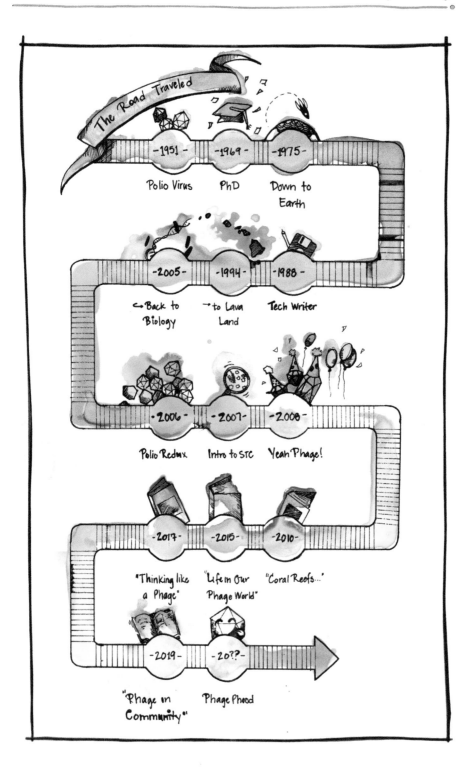

Merry Youle's Abridged Bio

1951   Shortly before the polio vaccine became available, this mere 7410 nt of ssRNA altered my life path. Nevertheless, recovery enabled me to hike, canoe, mix concrete, build rock walls, garden, and more for the next 55 years.

1969   Back when I completed my PhD in biology at Johns Hopkins University, phages warranted hardly a mention, being regarded mainly as tools for doing bacterial genetics.

1975   It was the time of communes, geodesic domes, underground newspapers, sandalwood incense, and whole-wheat bread—the perfect time for me to start Down to Earth, a natural foods restaurant in Oshkosh.

1988   The era of thriving software developers who distributed printed manuals with their products provided me with 20 years of free-lance work testing software and writing user documentation.

1994   A home at last—the start of decades in my lava tube house on the Big Island. Life was good off grid, but with Internet, a brisk climate, a year-round vegetable garden, and a cat.

2005   I outdid Rip van Winkle, having slept for more than 30 years in various pastures, unaware of the bustling activity in the biological field. I awoke to find microbiology had made stunning advances and the phages had been rescued from obscurity.

2007   When I happened upon Elio Schaechter's blog, *Small Things Considered*, I knocked on the door. Elio invited me in and generously encouraged me to share the helm and to contribute over 80 posts, the majority about the phage.

2008   No more software manuals! I began co-authoring research papers, review articles, and book chapters for the Forest Rohwer lab, and editing for other researchers.

2009   I sit at my desk, more incapacitated now by polio than when the virus first invaded, but continuing to write about the viral majority—my beloved phages.

2010   My first book opportunity came with an invitation from Forest to co-author his assessment of the role of microbes in the current plight of the coral holobionts: *Coral Reefs in the Microbial Seas*.

2015   Forest provided a second book project, this time as part of the team that produced his centennial phage book: *Life in Our Phage World*.

2017   One phage book led to another, this one being born from my own desire to tell more stories of phage wizardry: *Thinking Like a Phage*.

2019? Next to come will be *Phage in Community* to explore the collective role of Earth's most numerous, most diverse, and fastest evolving life forms.

???   Whether we love the phages or not, in the end—as Forest has noted—we're all phage phood.

CPSIA information can be obtained
at www.ICGtesting.com
Printed in the USA
BVOW11s1449071217

502211BV00023B/587/P

9 780990 494317